Springer Proceedings in Earth and Environmental Sciences

Series Editor

The series Springer Proceedings in Earth and Environmental Sciences publishes proceedings from scholarly meetings and workshops on all topics related to Environmental and Earth Sciences and related sciences. This series constitutes a comprehensive up-to-date source of reference on a field or subfield of relevance in Earth and Environmental Sciences. In addition to an overall evaluation of the interest, scientific quality, and timeliness of each proposal at the hands of the publisher, individual contributions are all refereed to the high quality standards of leading journals in the field. Thus, this series provides the research community with well-edited, authoritative reports on developments in the most exciting areas of environmental sciences, earth sciences and related fields.

More information about this series at http://www.springer.com/series/16067

Yaseen T. Mustafa · Sattar Sadkhan ·
Subhi Zebari · Karwan Jacksi
Editors

Recent Researches in Earth and Environmental Sciences

2nd International Conference on Advanced Science and Engineering 2019 (ICOASE2019) Zakho-Duhok, Kurdistan Region—Iraq, April 2–4, 2019

Springer

Editors
Yaseen T. Mustafa
University of Zakho
Zakho, Iraq

Subhi Zebari
Duhok Polytechnic University
Duhok, Iraq

Sattar Sadkhan
University of Babylon
Babil, Iraq

Karwan Jacksi
University of Zakho
Zakho, Iraq

ISSN 2524-342X ISSN 2524-3438 (electronic)
Springer Proceedings in Earth and Environmental Sciences
ISBN 978-3-030-18640-1 ISBN 978-3-030-18641-8 (eBook)
https://doi.org/10.1007/978-3-030-18641-8

This Springer imprint is published by the registered company Springer Nature Switzerland AG
The registered company address is: Gewerbestrasse 11, 6330 Cham, Switzerland

Preface

This book includes the papers presented during the 2nd International Conference on Advanced Science and Engineering 2019 (ICOASE2019) which was held in Duhok, Kurdistan Region, Iraq, in April 2019. This conference was jointly organized by two universities: University of Zakho and Duhok Polytechnic University. The theme of the book mainly focuses on the Natural and Environmental Sciences, and Earth Science and Geoscience as they are part of the ICOASE2019 tracks. It aimed to give a more concrete expression to the science and engineering applications with a new multilateral scientific forum that emphasizes the vulnerability and proactive remediation from an earth and environmental point of view.

The presented topics in both tracks of the ICOASE2019 were organized as:

- **Natural and Environmental Sciences**
  - Plant and Animal Ecology
  - Ecology and Pollution
  - Biodiversity
  - Soil Science and Geochemistry
  - Environmental Health and Waste Management
  - Environmental Biotechnology
  - Molecular Biology
  - Microbiology
  - Histology and Embryology
  - Biochemistry
  - Animal Physiology
  - Genetics
  - Entomology; Mycology and Parasitology
  - Pathogenic Bacteria and Immunology
  - Organic and Inorganic Chemistry
  - Analytical Chemistry

- Physical Chemistry
- Industrial Chemistry
- Quantum Chemistry

- **Earth Science and Geoscience**

  - Geology; Hydrology; and Geomorphology
  - Atmospheric Science
  - Human Geography
  - Photogrammetry
  - GIS and Remote Sensing Application
  - Forestry
  - Geodesy
  - Land Administration
  - Airborne and Terrestrial Laser Scanning
  - Hydrography
  - Cadastral Surveys
  - Spatial Data Infrastructure
  - Cartography
  - GNSS Application

More materials and information of the conference are available in https://icoase2019.uoz.edu.krd.

All submitted papers went through a rigorous (double-blind) peer review process commensurate with their tracks. In all, 39 research papers were submitted; each was reviewed by at least three reviewers, and only 12 were accepted (with 30.76% as acceptance rate). The process included full paper submission through a conference management system, the so-called EDAS. This system includes built-in software for similarity checking (iThenticate). The accepted similarity ratio is 25%. Next, the papers were assigned to reviewers through EDAS as they sent their review, comments, and recommendations through the system as well. The final decision was made by track chairs and editors after receiving revisions. Hence, a set of high-quality contributions to the recent researches in earth and environmental sciences were accepted.

We would like to thank everyone who contributed to this effort including paper authors, session presenters, reviewers, track chairs, program committee members, volunteers, and sponsors. Without their support, the event would not have been successful.

Zakho, Iraq — Yaseen T. Mustafa
Babil, Iraq — Sattar Sadkhan
Duhok, Iraq — Subhi Zebari
Zakho, Iraq — Karwan Jacksi
April 2019

Contents

Application of Computational Fluid Dynamics in the Simulation of Carbon Monoxide Distribution, a Case Study: Sayad Underground Tunnel in Tehran

Hazhir Karimi, Borhan Riazi and Mokhtar Mohhammadi

Abstract Vehicular emissions can easily contaminate the air quality of the traffic tunnel especially during traffic congestion or low vehicle speeds, which would pose serious health hazards to passengers and drivers. In this study, Carbon Monoxide (CO) concentration was simulated using a computational fluid dynamic inside the Sayad Tunnel in Tehran. Three-dimensional simulation using CFX is carried out to investigate the effect of tunnel geometry, vehicles' speed, the ventilation systems, and the fuel type on the CO concentration. For the making accurate simulation, the geometry of the tunnel was divided into grids and transmitted to the Ansys CFX. The result showed that the maximum concentration of CO for the top and bottom floors were 47.13 and 42.47 ppm respectively. Compared with the standards, these levels are not higher than the 8-h standard limit (109 ppm) for human health. In conclusion, using a three-dimensional CFD model for CO simulation is a suitable method to show and analyze the concentration in particular in the closed spaces.

Keywords CO · Numerical Simulation · CFD · Tehran

H. Karimi (✉)
Department of Environmental Science, Faculty of Science, University of Zakho, Kurdistan Region, Iraq
e-mail: hazhir.karimi25@gmail.com

B. Riazi
Department of Environmental Science, Islamic Azad University, Tehran Branch, Tehran, Iran
e-mail: briazi@pmz.ir

M. Mohhammadi
Department of Mechanical Engineering, College of Engineering, University of Tehran, Tehran, Iran
e-mail: mokhtar.m1370@gmail.com

Y. T. Mustafa et al. (eds.), *Recent Researches in Earth and Environmental Sciences*, Springer Proceedings in Earth and Environmental Sciences,
https://doi.org/10.1007/978-3-030-18641-8_1

1 Introduction

The simulation of the pollutant concentrations inside the closed spaces is an efficient way to explore the pollutant distributions through these spaces and assess the effects of vehicle speed, the locations of the fans, and the flow rate of the fresh air from fans. Simulating the concentration of the pollutants gives the better perspective to the place of the ventilation systems which could help the replacing contaminated indoor air with fresh air from outside, and it plays an essential role in the drivers' health and the efficiency of the operational expenses of construction costs [1].

Computational Fluid Dynamics (CFD) is one of the most proficient numerical methods that provide a qualitative and quantitative prediction of fluid flows using mathematical and numerical methods [2]. CFD gives an insight into flow patterns that are difficult, expensive or impossible to study using experimental techniques [3]. The advantages of the high resolution visualization, simulation for all desired quantities at a time, and application for actual flow domain have led to widespread use of CFD to study and predict various environmental phenomenon such as smokes, plumes, and [2, 4–7].

However, previous studies have focused on the smoke and heat distribution during the fire in which vehicles are stopped, and the effects of induced turbulence caused by vehicle motions has not been considered. Few studies have been caring out by recognizing vehicles movement at a slow speed, which may have a significant effect on pollutant distribution. For instance, Ashrafi et al. [3] used a computational fluid dynamics to the numerical simulation of airflow and carbon monoxide concentration inside the Resalat Tunnel in Tehran. Likewise, Dong et al. [8] investigated the CO_2 level in three traffic conditions (severe traffic congestion, traffic flow, and low vehicle speeds) in which fan conditions were considered to model the influence of mechanical winds on pollutant dispersion, and comparison with vehicular piston effect was also performed.

This study aims at the 3-D simulation of CO distribution in Sayad Tunnel of Tehran. The simulation of CO concentration was carried out using CFD based on the geometry of the tunnel, the consumed fuel, the type of the vehicle, and ventilation systems. The interaction between vehicular piston effect, ventilation and the volume of the input and output rates were also considered.

2 Methods

2.1 Tunnel Geometry

Sayad tunnel is the largest roadway tunnel in Iran with 10-km length. The geometry and overall layout of the tunnel is shown in Fig. 1. The tunnel is on two floors, one to go and the other to return the vehicles. Each floor has three lines of movement and a path for the emergency situations. The maximum depth of the tunnel is 32 m and

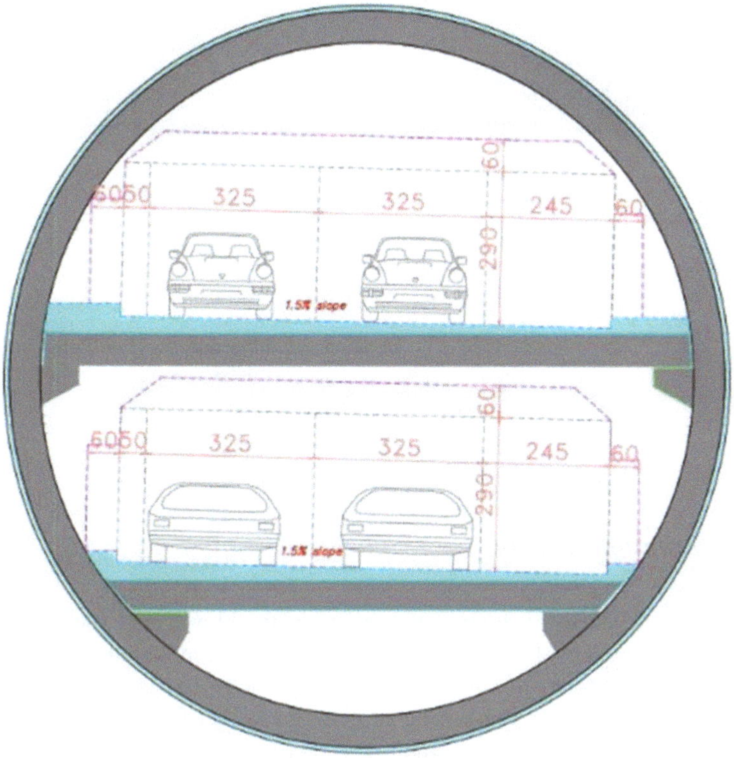

Fig. 1 The geometry of the tunnel

it has a 3.5 m height and a maximum slope of 5%. Each floor of the tunnel has 40 fans that are placed in 10 sections.

2.2 Vehicle Simulation

A vehicle with an average length and the medium volume of pollutants emission was considered for this study. Figure 2 shows the defined sample vehicle with a 4.8-m length. This vehicle is averaged in dimensions, velocity, and emissions. Two movement lines and one emergency line was considered with 10-m distances.

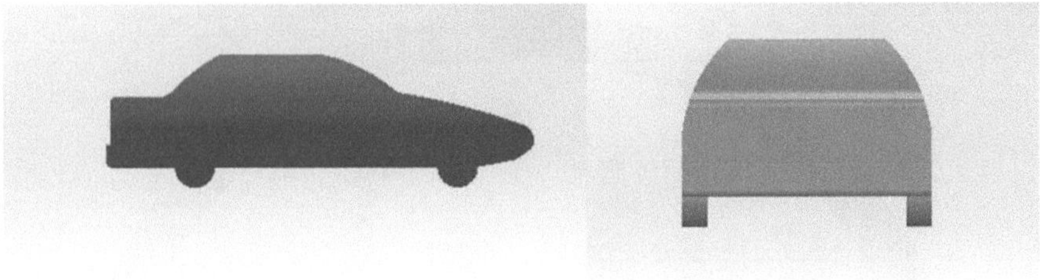

Fig. 2 Schematic presentation of a sample vehicle

Table 1 Characteristics of fans in different sections of the tunnel

Type	Location	Flow rate (m^3/s)	P (Pa)	Number
Blower	Upper floor	220.3	30.98	9
Suction	Lower floor	123.2	43.34	9

2.3 Ventilation System

The ventilation system of the tunnel consists of the blower and suction fans and jet fans. The volume of required air for the tunnel ventilation is estimated at 1983.04 m^3/s. Characteristics of jet fans are determined also based on the volume of the air and the pressure to transferring air between two jet-fan stations. The general system characteristic is shown in Table 1. According to the proposed model, nine fans with equal distance of 1000 m from each other are placed along the tunnel.

The momentum generated by the jet fan is also defined as Eq. 1:

$$V_m = V_d - V_o \ \& \ Thrust = Q \cdot V_m \cdot \rho \tag{1}$$

where Q is the passing volumetric flow rate through the jet fan, V_m is the pivot speed of the passing fluid that the jet fan can generate, ρ is the density of the fluid; V_o is the free flow speed around the jet fan and V_o is the velocity of the passing fluid through the fan. It is clear that if $V_d = V_o$, then the fan thrust it will be zero and practically the jet fan cannot create more speed.

2.4 Software Settings

For ease of grinding, the tunnel geometry is divided into several parts, and then network grids were imported into CFX. The input air from the entrance portal of the tunnel (in the presence of a piston effect in the movement of the cars) and the air supplied from the fans is the same as that of the urban environment air. To increase

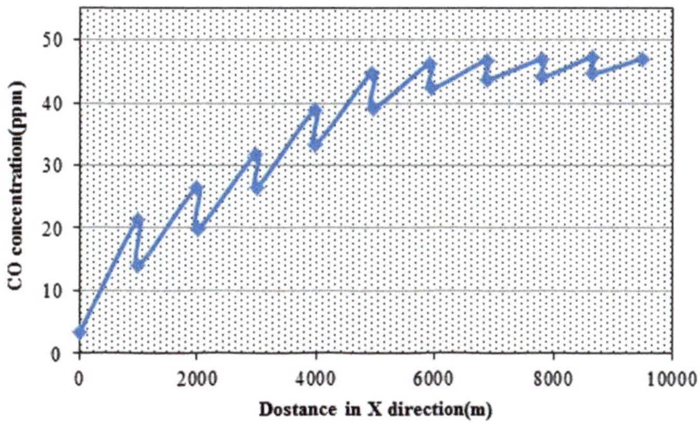

Fig. 3 Profiles of CO concentration in different sections inside the tunnel

the accuracy of simulating carbon monoxide concentration values over one year, a monthly average for air quality in these inputs is considered.

3 Results and Discussion

Simulation of CO pollutant was performed using CFD method and in CFX software. In the mode of cars motion, the tunnel is investigated according to its design conditions for two rows of cars at an average speed of 60 km/h. Figures 3 and 4 illustrate the average distribution of carbon monoxide inside the tunnel and the concentration contour of tunnel output respectively. In this case, the ventilation systems and piston effect were also considered. According to the graphs, the concentration of CO rose gradually during the tunnel and the CO_2 concentration reached about 47 ppm at the output, which was also under a hazardous level for the environment.

Figure 5 illustrates the standard level of CO in different exposure time. According to the figure the concentration of 50 ppm for 3 h is not at the hazardous level, and given that the speed of the vehicles is considered 60 km/h and the tunnel length of 10 km, the total duration of the time that the vehicles run through the tunnel will be 10 min. Thus there will be no concern for the health of people, and the ventilation system will respond to the dilution of the air. It is necessary to note that the best concentration condition of 35 ppm for 8 h of inhalation is considered and breathing carbon monoxide 9 ppm for a long time is mentioned permissible.

Use of the CFD model as a 3D model in the contaminants emission simulation showed that this model could be able to simulate the pollutant contamination inside the tunnels. In similar studies by Ashrafi et al. [3] and Dong et al. [8], they investigated the CO_2 level in traffic tunnel using CFD. They resulted that CFD is a powerful tool for simulating CO concentration in the tunnel. They also found that the vehicular

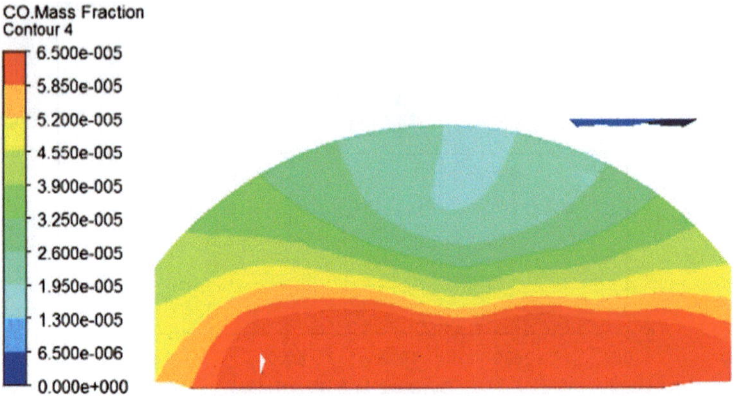

Fig. 4 Concentration of the contour in the tunnel

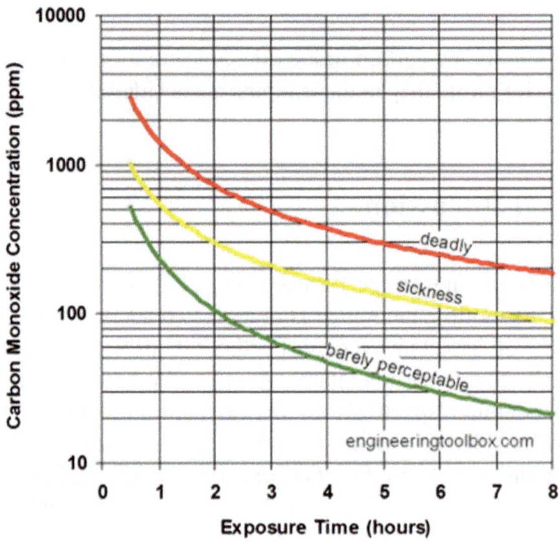

Fig. 5 Standard level of CO

piston effect could efficiently remove vehicular emissions when vehicles travel is at higher speed. The result of this study agrees with those conducting by Ashrafi et al. [3] and Dong et al. [8].

4 Conclusion

In this study, CO concentration in Sayad Tunnel was simulated using CFD that is a 3D numerical method. Simulation of CO contaminant showed that this model is a power model to simulate CO distribution. The result indicated that piston effect and ventilation system helped the elimination of emissions adequately under a vehicle speed. Distribution of CO concentrations depends on several factors, including tunnel geometry, vehicle movement, fan locations, and fresh air flow of the fans. In the lots of studies, the movement of vehicles has not been considered and the actual concentration of CO has not been achieved. However, we simulated the turbulent boundary conditions and the actual CO concentration considering the speed and movement of vehicles.

One of the important points in simulation and modelling is validation and accuracy assessment. This study was conducted at the time of designing. Therefore, it is not possible to getting air contamination samples during the use of the tunnel to compare the level of the CO with the simulated results of this study. It is strongly suggested that in the future studies, researches will compare the results with the level of the pollutants by sampling in different sections and profiles of the tunnel.

Acknowledgements The authors would like to thank Tehran Municipality for their financial support.

References

1. Lee, S.R., Ryou, H.S.: A numerical study on smoke movement in longitudinal ventilation tunnel fires for different aspect ratio. Build. Environ. **41**(6), 719–725 (2006)
2. Eftekharian, E., Abouali, O., Ahmadi, G.: An improved correlation for pressure drop in a tunnel under traffic jam using CFD. J. Wind Eng. Ind. Aerodyn. **143**, 34–41 (2015)
3. Ashrafi, K., Shafie-pour, M., Kalhor, M., Esfahanian, V.: Numerical simulation of air pollutant distribution in urban tunnels. Environ. Model. Assess. **17**(5), 555–564 (2012)
4. Neofytou, P., Venetsanos, A.G., Vlachogiannis, D., Bartzis, J.G., Scaperdas, A.: CFD simulations of the wind environment around an airport terminal building. Environ. Model Softw. **21**(4), 520–524 (2006)
5. Clausen, P.A., Liu, Z., Xu, Y., Kofoed-Sørensen, V., Little, J.C.: Influence of air flow rate on emission of DEHP from vinyl flooring in the emission cell FLEC: measurements and CFD simulation. Atmos. Environ. **44**(23), 2760–2766 (2010)
6. Hanna, S.R., Hansen, O.R., Ichard, M., Strimaitis, D.: CFD model simulation of dispersion from chlorine railcar releases in industrial and urban areas. Atmos. Environ. **43**(2), 262–270 (2009)
7. Karim, A., Nolan, P.: Modelling reacting localized air pollution using computational fluid dynamics (CFD). Atmos. Environ. **45**(4), 889–895 (2011)
8. Dong, J., Tao, Y., Xiao, Y., Tu, J.: Numerical simulation of pollutant dispersion in urban roadway tunnels. J. Comput. Multiph. Flows **9**(1), 26–31 (2017)

Dendroclimatological Analysis of *Pinus brutia* Ten. Grown in Swaratoka, Kurdistan Region—Iraq

Tariq K. Salih, Muzahim Saeed Younis and Salih T. Wali

Abstract It is well known that there is a strong relationship between the amount of precipitation and radial growth of the trees, but it is not known up to which extent they depends on each other. On the other hand, the data about the amount of precipitation is available from 1976 in Duhok governorate in general. One of the problems studied here is to estimate the quantity of the precipitation using the width of annual ring for the periods prior to 1976, in order to see the trend of rain and snow fall in the region. The data used in this study came from 387 sample pairs of precipitation and diameter growth. These data were undergone data processing for the purpose of developing of regression equations between the width of diameter growth as dependent variable and the amount of yearly precipitation as independent variables in regression equations. Accordingly, thirteen regression equations were developed; two of them were simple regression equations, one polynomial, and the rest of equations were nonlinear. These equations were undergone several measures of precision for the purpose of selecting the most appropriate one which fits our data set. Ultimately the non-linear regression equation: $D_g = 0.293781 + 0.0000371 \times p^{1.40698}$ was selected, with an adjusted R^2 of 76.22, standard error of estimate of 0.1283 and DW of 1.88 The selected equation can be used to estimate the amount of radial growth of a tree in the region by substituting the amount of precipitation in the equation.

Keywords Annual ring · Dendrochronology · Precipitation · Increment borer · Diameter growth

---

T. K. Salih (✉) · M. S. Younis
College of Agriculture, University of Duhok, Duhok, Kurdistan Region, Iraq
e-mail: Tariq.k.salih@hotmail.se

M. S. Younis
e-mail: muzahimsaeedyounis@gmail.com

S. T. Wali
College of Agriculture & Forestry, University of Mosul, Mosul, Iraq
e-mail: stwali@uod.ac

Y. T. Mustafa et al. (eds.), *Recent Researches in Earth and Environmental Sciences*, Springer Proceedings in Earth and Environmental Sciences,
https://doi.org/10.1007/978-3-030-18641-8_2

1 Introduction

Pinus brutia Ten is the native coniferous tree species grown in Kurdistan of Iraq. It covers up to 50,000 ha [1], which is considered as a 10% of the overall forest area in the region. The rest of the area is covered by broadleaves, particularly the oak trees species [2–4].

The annual rings are more conspicuous in the cross sectional area and, also on the extracted core from brutia pine as compared with the other naturally grown trees in the region [5]. Furthermore, the process of cores extraction from pine trees via increment borer is much easier than extraction from oak trees. For such reasons this species was used for such dendro-climatological study.

The relationship between climate and tree-ring width is a part of so-called Dendrochronology.

The main reason behind using the diameter growth in this chronological study is the variation of growth rate in different seasons within one year. Growth is the increase of the size of tree under a period of time [6–8]. The growth takes place in different parts of a tree at the same time. It can be expressed in different parameters, and therefore there is diameter growth, height growth, and volume growth [8]. Growth is the product of an interaction of several abiotic and biotic factors, over a tree and forest, through the time. Diameter growth is the difference between the diameters of a tree under two successive years [9].

In measuring the width of annual rings, one may face some problems, the difficulties in distinguishing the true boundaries of annual rings and mistakes which may arise in counting the annual rings [10].

Leonardo da Vinci [11] realized the relation between climate and width of tree rings. He is considered, as the father of dendrochronology [12]. However, the beginning of broad scientific knowledge of wood formation, tree rings and climate, can be dated to the middle of the nineteenth century, when Theodor Hartig and others postulated their theory of tree-ring formation as a consequence of low temperature in winter [13]. The nature can be considered as a book of many pages and each page tells us a fascinating story [14].

The climatic variation in a location can be read on the annual rings of a tree from the same location, because the width of these rings has a high relationship with the climatic conditions [15]. This characteristic can be used to study the previous climate variation, in order to see the trend of some aspects of climate in the past [16, 17].

This study is partially dealing with the possibility of anticipating the past precipitation of the region using the annual growth of the trees. The relationship between the width of annual rings and the precipitation has been studied intensively by many researchers such as [18–20].

Although the chronological investigations have been conducted throughout the world since the beginning of the last century, in our country this attempt can be considered as the pioneer one at least for the mountainous region of Kurdistan. The tree rings can be used as a record containing year by year something of the climate history in the annual rings.

It seems that the growth is directly correlated with precipitation and inversely correlated with the temperature [21]. If the temperature increases, tree growth may decrease, depending on whether the heat increase is accompanied by an adequate amount of precipitation. Tree growth is lowest when there are conditions of high temperatures and low precipitation, while it is greater when temperature is lower and moisture is higher [22].

The tree-ring width have been used in different purpose, such as reconstruction of records of past climatic changes [16, 23, 24]. The characteristics of the dated rings can be used to reconstruct past variations in drought [25, 26]. Mäkinen and Vanninen [27] found that there were no effect of temperature on wood production, but a marked correlation between the precipitation and variation in ring width.

The recording of climatic data in Duhok governorate was started in 1976, which means that such data is not available before that date. Therefore, this study aimed at investigating of the following: (1) to see up to which extent the annual growth of the tree is accompanied with the precipitation. (2) To investigate the possibility of predicting the past precipitation of the study region at any year by measuring the tree-ring width at that year.

This paper seeks to address the following questions:

1. To determine the relationship between the radial growth and the precipitation.
2. To regress the annual ring width with the annual precipitation.
3. To regress the annual precipitation with annual radial growth to be used for estimating the annual precipitation for the studied area in the past time.
4. To estimate the effect of precipitation on the size of radial growth of brutia pine trees.
5. The knowledge gained from this study will help with the reconstruction of past precipitation using tree-ring records of extremely long-lived pine trees.

2 Materials and Methods

2.1 Site and Location Information

The data used in this study came from a sample of pine trees collected from Swaratoka region. Swaratoka is mountainous area which lies at about 1250 m above sea level. The mean annual precipitation is 800 mm. The data of precipitation were obtained from the directory of meteorology office in Duhok. The latitude and longitude for the region are 370.545 and 4313.691 respectively.

The subjectively selected sample was consisted of 12 open grown, undamaged, well formed, healthy, and vigor trees. This sample of trees were cored using an increment borer at a height of 1.3 m above the ground level, because there is a significant correlation between the radial growth and the precipitation at such height [28].

2.2 Laboratory Work and Core Preparations

The lab and office work included smoothing of the extracted increment core for the sake of getting a better appearance of clean annual rings. Furthermore a handheld sandpaper was used for getting a readable surface of such cores.

These increment cores were fastened on a woody basement plate for photographing. The images were taken with a high resolution Camera (Nikon digital Camera D3100). These images were transferred to the computer, to be scanned with computer scanned equipment called, CDendro and Co-Recorder, which is specially designed for image analysis and measuring of annual rings. Using this system, the width of each annual ring along each diameter-based radius is obtained. Accordingly 387 observation of diameter, age and diameter growth were obtained. These data were undergone both mathematical and statistical analysis. For such purposes, the Excel, Stratigraphic plus 4 and SAS were used for data processing, developing equation, and testing the performance of the developed equations to select the most appropriate one for estimation of diameter growth corresponding to different amount of precipitation.

From each selected tree, in the sample, between 20 and 40 observations were taken.

The reason of taking few numbers of trees in a sample is:

1. If the same tree is used in taking the required measurements, this ensures keeping (all factors affecting the growth) as constant except precipitation, which was the studied explanatory variable.
2. The tree, which is living in a location has the same site factors, and has the same density, because our selected trees was from open area, which means that they were free from inter and intra-competition.
3. It was not easy to find many trees having the same aspect, same altitude.

2.3 Procedure of Achieving the Diameter and Diameter Growth

1. Transferring the image of cores to the computer.
2. Using the CDendro and Co-Recorder, the borders of annual rings were determined and accordingly the number of rings was accounted to be used for determination of the age of the tree.
3. Starting from the pith of the diameter of the first ring was considered as the first diameter and the twice of the distance between the first ring and second ring was taken as the diameter growth for the first ring.
4. The next diameter was the first one plus the diameter growth of the first ring, i.e. $D2 = D1 + Dg1$, and the double of distance between the second and third ring was taken as $Dg2$ and so on (see Fig. 1).

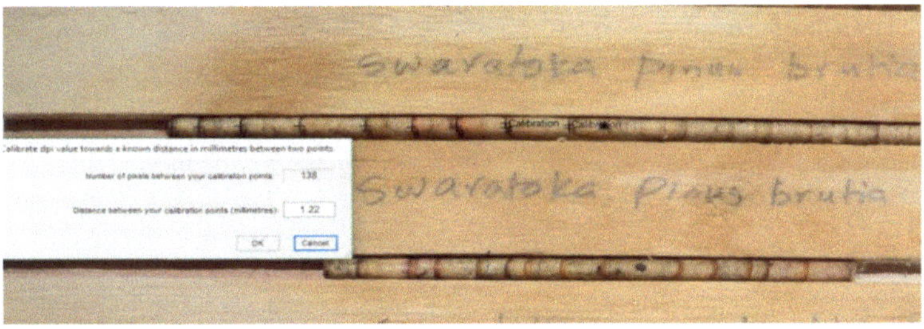

Fig. 1 Shows diameter and diameter growth calibration using co-recorder package

Table 1 Shows a summary of the data used in this study

Variable	Mean	SD	Min.	Max.
Diameter	18.184	10.61	0.200	44.80
Precipitation	779	297.03	200.7	1402
Diameter growth	0.7464	0.2626	0.24	1.35

5. The year belonging to each diameter growth was calculated, using simple arithmetic, for example if the total numbers of annual rings at DBH is 30 rings and the cores were taken in 2017, this means that the first ring belongs to (2017 − 30) = 1987.
6. After the determination of the year at which each annual ring was formed, the precipitation was obtained from the directory of meteorology office in Duhok, belonging to the same region.
7. In such cases, the required data became ready for analysis and development of regression equations (Table 1).

3 Results and Discussion

3.1 Development of Equations

At the beginning a scatter plot were done for our data to see the trend of relationship, in order to determine the type of models to be tested (see Fig. 2).

The plotted points show some types of non-linear relationships, which in turn declare some outlier within data. This fact was verified when the first regression equation was developed.

By excluding the outliers from the original data, the R'^2 increased from 0.55 to 0.75. Table 2 shows the developed equation after excluding the 19 abnormal observations from the total of 387. Using the available computer package; namely Stratigraphic plus 4, SAS and Excel and different forms of linear and nonlinear regression equations were adopted and tested to see how well they fit our data.

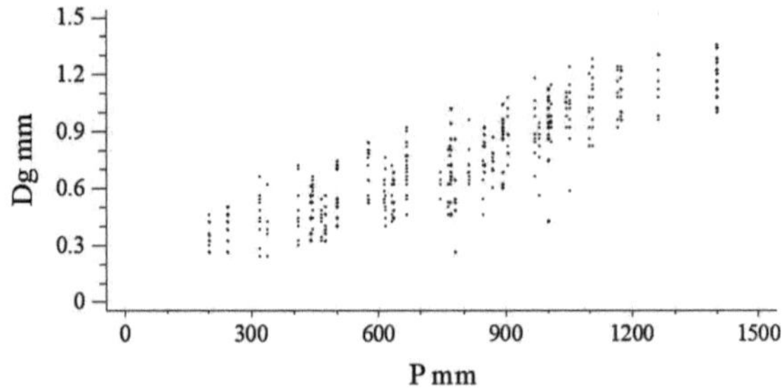

Fig. 2 Shows that the trend of relationship between diameter growth and the amount of annual precipitation

Table 2 Shows the developed regression equations along with some measures of precision

No.	Equations	R^2	SEE
1	$Dg = 0.146496 + 7.66968\text{E}{-4} * P$	75.65	0.1299
2	$\ln Dg = -4.94356 + 0.69769 \ln D$	69.14	0.2122
3	$Dg = 0.252163 + 4.497\text{E}{-4} * P + 2.06585\text{E}{-7} * P^2$	76.11	0.1285
4	$Dg = 0.00355152 * P * e^{0.804555}$	74.62	0.1325
5	$Dg = 0.293781 + 3.71\text{E}{-5} * P^{1.40698}$	76.22	0.1283
6	$Dg = 5.545\left(1 - 0.978858 * e^{(-0.000159428P)}\right)$	75.12	0.1312
7	$Dg = 1.58549 * e^{\left(\frac{-540.4}{P}\right)}$	65.81	0.1538
8	$Dg = 4.72837 * (1 - 0.331808 - (0.00050595 * P))^{7.37035}$	76.14	0.1285
9	$Dg = 1456.61 * (1 - 0.784949e^{(-0.0740491 * \ln P)^{11.5793}})$	73.65	0.1350
10	$Dg = 0.0069951 * \left(1 + e^{(3.82133+0.00102032 * P)}\right)$	75.22	0.1309
11	$Dg = 3.53654 * e^{(-e^{(0.988845-0.00068458 * P)})}$	76.23	0.1282
12	$Dg = 62.012 * P^{(-1.07061)} + 0.000877 * P$	76.35	0.1279
13	$Dg = \frac{1.94732}{(1+6.64125 * e^{(-0.00177669 * P)})}$	76.44	0.1277

where
Dg = diameter growth (mm)
P = precipitation (mm)

3.2 *Measures of Precision and Biased*

The following measures of precision were used for comparing the precision of the developed regression equations:

1. Adjusted coefficient of determination (R'^2)

This measure is calculated according to the following:

$$R'^2 = \left(1 - \frac{\frac{regression\,ss}{k}}{\frac{Total\,ss}{n-1}}\right) \quad (1)$$

2. Standard error of estimate (SEE), which can be calculated using the following formula:

$$\mathrm{SEE} = \sqrt{\frac{Residual\,ss}{n-k-1}} \tag{2}$$

3. Ohtomo's unbiased test:

$$\hat{y} = b + my \tag{3}$$

This actually is a simple regression equation between the estimated values of dependent variable from the selected equation and the actual values of the dependent variable where:

k = number of independent variables in the equation
n = the number of observation
b = y-intercept of Ohtomo's model
m = regression coefficient for Ohtomo's model.

3.3 Selection Procedure

Table 1 shows that the dependent variable of the developed equations appeared in two different forms; in the second one, it appeared in logarithmic form, while in all others equations in original form. Therefore the performance of the second one can't be tested with the rest of equations [29]. The logarithmic function was remained as a candidate to be compared with other equations, using other measures of precision such as [29, 30]. Meanwhile, also that the adjusted R'^2 ranged from 65.81 to 76.44 and their standard error of estimate ranged from 0.1277 to 0.1538.

It is well known that the precision of an equation is directly proportional with it's R'^2 and indirectly proportional with SEE, therefore, No. 7 of Table 2 was dropped from the list of competing functions because of having lowest R^2 and highest SEE.

Out of the twelve regression equations, one is simple linear equation with an R'^2 of 75.65 and SEE of 0.1299, and a polynomial with R'^2 of 76.11 and of an SEE of 0.1285.

The rest of models are non-linear equations. Their R'^2 lies between 73.65 and 76.44, and their SEE lie between 0.1277 and 0.1350. These twelve equations were undergone another test called Ohtomo's unbiased test for testing of precision of the equations. This test was originally developed for testing the accuracy of tree volume equations, but it can be used in other circumstances with some modifications. The starting point of this test is relating the estimated values of the dependent variable will be regressed on its actual values in a simple linear regression equation, and as follow:

Table 3 Shows the Ohtomo's test for the remaining candidates

No.	Equation models	R^2	SEE	M	b	b + (1 − m)
1	$Dg1 = b + mDg$	75.65	0.113	0.7565	0.1817	0.4252
2	$Dg2 = b + mDg$	74.14	0.1036	0.6661	0.2326	0.5666
3	$Dg3 = b + mDg$	74.78	0.1182	0.7731	0.1669	0.3938
4	$Dg4 = b + mDg$	76.35	0.1121	0.7646	0.1763	0.4117
5	$Dg5 = b + mDg$	75.18	0.1314	0.8682	0.0852	0.2170
8	Dg7 = b + mDg	76.03	0.1619	1.095	0.126	0.2214
9	Dg8 = b + mDg	74.00	0.1200	0.7687	0.1695	0.4008
10	$Dg9 = b + mDg$	75.39	0.1113	0.7395	0.1955	0.456
11	$Dg10 = b + mDg$	76.36	0.1121	0.7647	0.1738	0.4091
12	$Dg11 = b + mDg$	76.48	0.1110	0.7599	0.1795	0.4196
13	$Dg12 = b + mDg$	76.57	0.1157	0.7705	0.1708	0.4004

$$Y = b + mY \tag{4}$$

The most accurate equation is the one, which has estimated values of: $\hat{Y}$ almost equal to the corresponding actual values, (Y). This situation meets when $b = 0$ and $m = 1$ (Table 3).

As it can be seen that the R'^2 and SEE ranged between 74–76.57 and 0.1036–0.1314 respectively, which means that their efficiency in estimating the dependent variable is very close to each other. In the view of Ohtomo's test, the values of y-intercept ranged from −0.1264 to 0.2326 and the regression coefficient ranged between 0.6661 and 1.095. Taking both (b and m) criteria into consideration, the following index was proposed to see the relative efficiency of the competed equations in meeting the Ohtomo's criteria:

$$Proposed\ Index = b + (1 - m).$$

The first term shows the absolute deviation of n from zero, and the second one calculates the absolute deviation of m from one. The proposed index calculates the sum of the deviation of both (b) and (m) from zero and one respectively. The candidate equation, which has the lowest value of the index, will be considered as the best one. Accordingly, the fifth equation was finally selected for predicting the diameter growth corresponding to the known value of precipitation, because of being the best one as compared with the others. It is worth to mention that the selected equation has no autocorrelation because, of having a value of 1.88 of Durbin Watson.

One of the most important property of the selected equation is the independency of the residuals. $\sum (y - \hat{y})^2$ with the values of the independent variable. In other word there should be no trend for the plotted point of residuals with X-values, which is precipitation in this study (see Fig. 3). The residuals must be normally and independently distributed, with (a mean of zero and variance of 1).

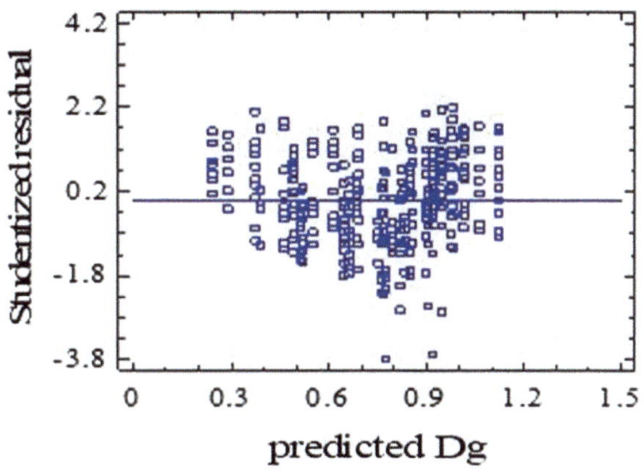

Fig. 3 Shows that the residuals are scattered, are homogenous throughout the range of our dataset

$$\text{Residuals} \sim NID\left(0, \sigma^2\right) \tag{5}$$

This means that the selected equation has the same accuracy in estimation for the whole range of data.

The selected equation was finally solved for relating of the precipitation with annual radial growth, as follow:

$$\mathrm{P} = \left(\frac{\mathrm{D_g} - 0.293781}{0.0000371}\right)^{0.710741} \tag{6}$$

The last equation can be used for predicting the precipitation corresponding to the known width of annual ring growth.

4 Conclusions and Recommendations

At the beginning both oak trees and pine trees were tested for such study, but the appearance of annual rings on the extracted cores from the pine trees was much more conspicuous as compared with the cores extracted from the oak trees. Moreover, the process of getting cores via increment borer in pine trees was much easier than oak trees, therefore it is recommended to use pine trees for such studies in the future.

The total sum of squares of deviations of the original values of the diameter growth from the mean $\sum(Y - \bar{y})^2$ depends on many factors and not only the yearly precipitation. The precipitation explained almost 75% of the total variation, which is considered as very high. This high value of the adjusted R'^2 is due to two mean reasons: the first one is the method of taking the samples, as mention before. With such method, we tried to minimize the other factors that have a considerable effect on the diameter growth. The second one is the excluding of the 19 outliers from our dataset.

References

1. Shahbaz, S.: Trees and shrubs, a field guide to the trees and shrubs of Kurdistan region of Iraq. Univ. Duhok. (2010)
2. Chapman, G.: Notes on forestry in Iraq. Emp. Forest. Rev. 132–135 (1950)
3. Şefik, Y.: Forests of Iraq. J. Fac. Forest. Istanbul Univ. **31**(1) (1981)
4. Nasser, M.: Forests and forestry in Iraq: prospects and limitations. Commonw. Forest. Rev. 299–304 (1984)
5. Therrell, M.D., Stahle, D.W., Ries, L.P., Shugart, H.H.: Tree-ring reconstructed rainfall variability in Zimbabwe. Clim. Dyn. **26**(7–8), 677 (2006)
6. Husch, B., Beers, T.W., Kershaw Jr., J.A.: Forest Mensuration. Wiley (2002)
7. Nord-Larsen, T.: Modeling individual-tree growth from data with highly irregular measurement intervals. For. Sci. **52**(2), 198–208 (2006)
8. Philip, M.S.: Measuring Trees and Forests. CAB international (1994)
9. Van Laar, A., Akça, A.: Forest Mensuration, vol. 13. Springer Science & Business Media (2007)
10. Palakit, K., Siripattanadilok, S., Duangsathaporn, K.: False ring occurrences and their identification in teak (*Tectona grandis*) in north-eastern Thailand. J. Trop. For. Sci. 387–398 (2012)
11. Haneca, K.: Tree-Ring Analyses of European Oak: Implementation and Relevance in (Pre-)historical Research in Flanders. Ghent University (2005)
12. Cook, E.R.: Dendrochronology and Dendroclimatology. Encycl. Environmetrics **2** (2006)
13. Studhalter, R.: Tree growth. Bot. Rev. **21**(1–3), 1 (1955)
14. Douglass, A.E.: Some aspects of the use of the annual rings of trees in climatic study. Sci. Mon. **15**(1), 5–21 (1922)
15. Jacoby, G.C., D'Arrigo, R.D.: Tree ring width and density evidence of climatic and potential forest change in Alaska. Global Biogeochem. Cycles **9**(2), 227–234 (1995)
16. Fritts, H.: Tree Rings and Climate. Academic Press, New York (1976)
17. Schweingruber, F., Fritts, H., Braker, O., Drew, L., Schar, E.: The X-ray technique as applied to dendroclimatology. Tree Ring Bull. (1978)
18. Brubaker, L.B.: Spatial patterns of tree growth anomalies in the Pacific Northwest. Ecology **61**(4), 798–807 (1980)
19. Graumlich, L.J.: Response of tree growth to climatic variation in the mixed conifer and deciduous forests of the upper Great Lakes region. Can. J. For. Res. **23**(2), 133–143 (1993)
20. Watson, E., Luckman, B.H.: The dendroclimatic signal in Douglas-fir and ponderosa pine tree-ring chronologies from the southern Canadian Cordillera. Can. J. For. Res. **32**(10), 1858–1874 (2002)
21. Fritts, H.: The relation of growth ring widths in American beech and white oak to variations in climate. Tree-Ring Bull. **25**, 2–10 (1962)
22. Allen, C.D., Macalady, A.K., Chenchouni, H., Bachelet, D., McDowell, N., Vennetier, M., Kitzberger, T., Rigling, A., Breshears, D.D., Hogg, E.T.: A global overview of drought and heat-induced tree mortality reveals emerging climate change risks for forests. For. Ecol. Manage. **259**(4), 660–684 (2010)
23. Douglass, A.E.: A method of estimating rainfall by the growth of trees. Bull. Am. Geogr. Soc. **46**(5), 321–335 (1914)
24. Michaelsen, J., Haston, L., Davis, F.W.: 400 years of Central California precipitation variability reconstructed from tree-rings. JAWRA J. Am. Water Res. Assoc. **23**(5), 809–818 (1987)

25. Blasing, T., Stahle, D., Duvick, D.: Tree ring-based reconstruction of annual precipitation in the south-central United States from 1750 to 1980. Water Resour. Res. **24**(1), 163–171 (1988)
26. Fritts, H.C., Lofgren, G.R., Gordon, G.A.: Variations in climate since 1602 as reconstructed from tree rings. Quatern. Res. **12**(1), 18–46 (1979)
27. Mäkinen, H., Vanninen, P.: Effect of sample selection on the environmental signal derived from tree-ring series. For. Ecol. Manage. **113**(1), 83–89 (1999)
28. Diller, O.D.: The relation of temperature and precipitation to the growth of beech in northern Indiana. Ecology **16**(1), 72–81 (1935)
29. Furnival, G.M.: An index for comparing equations used in constructing volume tables. For. Sci. **7**(4), 337–341 (1961)
30. Ohtomo, E.: A study on preparation of volume tables. J. Jpn. For. Soc. **38**(5) (1956)

Risk Assessment of Dangerous Natural Processes and Phenomena in Mining Operations

Elena Kulikova

Abstract One of the main factors that characterize the conditions of mining operations in the mines of the North of Russia is a sharp increase in the disturbance of ores and rocks in the exposed areas of minefields with increasing depth of development. The presence of permafrost, salt layers, aggressive and pressure groundwater, strongly fractured and prone to swelling rocks, significant tectonic disturbance of rock mass can lead to deformation of the lining. All this is associated with the emergence and development of geological and, above all, geodynamic risks. The main indicators of geological risk assessment, allowing to conduct a reasonable comparative analysis of possible emergencies and measures to prevent them, are differentiated and integrated characteristics of the specific economic and individual risk of loss for 1 year. Full values of such risks are used as additional indicators. The results of the geological risk assessment are the basis for determining the need, composition, sequence of implementation and socio-economic efficiency of measures to prevent natural emergencies caused by the development of geological hazards, as part of the project documentation. The paper is aimed on the investigation of geodynamic danger zone around ultra-deep shaft; the level of risk of its structural failure; the main parameters, control of which should be required in the construction of deep mine shafts in the permafrost and tectonic instability; monitoring of emergency situations during the whole period of the shaft existence.

Keywords Underground construction · Risk · Hazard · Monitoring · Mining operation

E. Kulikova (✉)
Department "Construction of Underground Structures and Mining Enterprises", National University of Science and Technology MISIS, Moscow, Russia
e-mail: fragrante@mail.ru

Y. T. Mustafa et al. (eds.), *Recent Researches in Earth and Environmental Sciences*, Springer Proceedings in Earth and Environmental Sciences,
https://doi.org/10.1007/978-3-030-18641-8_3

1 Introduction

Mining activities are constantly accompanied by additional specific risk, the existence of which is associated with the fact that mining is always carried out not only in conditions of incomplete knowledge about the subsoil, but also with the continuous movement of mine workings in their space.

The basis of all projects and plans for the development of mining operations are not the actual data on the subsoil, but only geological models of it, which objectively have one or another level of error, and unstable at various points. During the use of geological information, errors are transformed into errors of technological, investment and other solutions. The background level of geological errors leads to a decrease in the expected technical and economic indicators of construction.

Therefore, risk assessment is an essential element of pre-investment analysis in mining.

The study deals with the construction of an ultra-deep shaft (over −2000 m) at the mine in Norilsk. The area (Fig. 1) is characterized by natural landscapes, formed as result of complex geological and climatic processes, adjacent to the territory, which has a high degree of industrial development and production and processing of large Cu-Ni deposits. This neighborhood determines the degree of change in primary natural complexes as result of the impact on the environment of various adverse factors accompanying industrial production.

The climate of the region is severe, sharply continental, subarctic. The average annual temperature is −9.8 °C. In the coldest period (January–February), the temperature drops to −50 °C and below, in the hottest (second half of July)—reaches +30.5 °C and above. The humidity is relatively low: the annual rainfall is less than

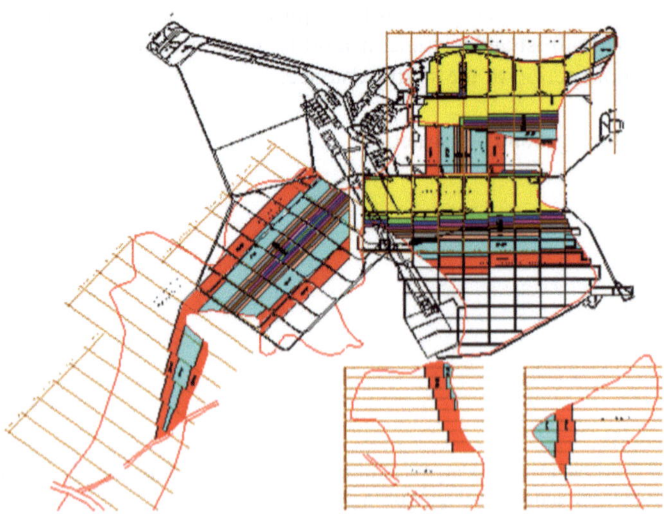

Fig. 1 The area of mining operation

400–500 mm. The summer is intermittent short. Snow cover lasts from late September to mid-June. Winter is frosty, with frequent and long purges (up to 200 days a year).

The rock mass is composed of dolomites, dolomite marls, anhydrites. There are low-power horizons of sandstone lenses. There is a violation of ores and rocks in the exposed areas of minefields with increasing depth of development. In the thickness crossed by the shaft there are permafrost, salt layers, aggressive and pressure groundwater, strongly fractured and prone to swelling of the rock. There is also a significant tectonic disturbance of the rock mass. At depths of 141.1–147 m and 148.1–149.8 m there are zones of rock weakening. At a depth of 945–2000 m there is a hydrogeological unstable zone.

The main goal is to define the level of geodynamic risk in zones of disturbance, to forecast the probability of water breakthrough of water into the shaft and to propose algorithm of monitoring of risk development during operational an exploitation works.

2 Groups of Mining Risks

In the described conditions, mining risk can be divided into the following main groups [1–9]:

- geological risk, determined by the accuracy of estimates of construction conditions;
- legal risk, due to problems associated with the implementation of existing legal requirements and the possibility of eliminating existing violations of existing license agreements;
- investment risk, related to the correct assessment of the necessary investment costs for construction;
- operational risk, due to the possibility of achieving and maintaining the required economic parameters (cost, volume of construction) in the construction of mining facilities.

Estimates of the importance of individual components of mining risk made for this study are presented in Table 1.

All decisions about the development of the site to accommodate shafts and other workings have to be accepted from the point of view not to exceed the limits of harmful impact on the natural environment. It is difficult to establish these limits, since the thresholds for the impact of many construction and natural factors are practically unknown. Therefore, environmental risk calculations in this case can only be probabilistic and carried out separately for human health and the environment. The state environmental expertise, when approving the feasibility study of construction projects, placement of facilities, etc., verifies the provision of acceptable environmental risk and its guarantee in them.

Table 1 Estimates of the importance of individual components of mining risk

Type of risk	Rank reflecting the degree of importance of the risk component, %
Geological risk	100.0
Material prices	33.3
Operating costs	33.3
Location	23.8
Capital expenditure	22.2
Management	20.6
Registration of rights	20.6
The tax regime	15.9
Geological features	11.1
The technology	6.3
Construction stage	4.8
The prospect of increasing the volume of construction works	3.2

2.1 Environmental Risk

Environmental risk at this facility should be assessed at all stages of the shaft existence and can be manifested in the form of intensive gas-dust emission of substances, excess noise and vibration impact on the territory of the mining branch, as well as unacceptable soil sediment, shaft flooding, development of deformation processes, etc. [10].

When assessing environmental risk in this case, the following factors have to be taken into account:

- Geological—the state of the geological environment, i.e. the thickness of soils used for underground construction (composition and properties of soils, groundwater and their regime, dangerous geological processes, etc.);
- Technological—the composition of works carried out in the construction of the shaft (dewatering, soil consolidation, etc.), their impact on the environment;
- Constructive—physical–mechanical and other properties of building materials and structures (strength, deformability, corrosion resistance, etc.).

2.2 Geological Risk

Geological risk assessment is performed on the basis of engineering survey materials. At the same time, in each case it is necessary to consider and evaluate at least two forecast scenarios for the development of natural emergencies with the most probable and maximum possible losses (risks).

As the main final indicators of the geological risk assessment, which allows passing to the reasonable comparative analysis of possible emergencies and actions for their prevention, one can use the differentiated and integrated characteristics of specific economic and individual risk of losses for 1 year, and as additional indicators—full values of such risks. The results of the geological risk assessment are the basis for determining the need, composition, sequence of implementation and socio-economic efficiency of measures to prevent natural emergencies caused by the development of geological hazards, as part of the project documentation [5].

The process of assessing the level of geological risk in our investigation begins with the identification of geological hazards. The purpose of identification is to establish the following parameters: characteristics, indicators, conditions, factors and patterns of development of all existing manifestations of geological hazards in the estimated area. Including the definition of their distribution, the scope of the geological environment, genesis, age, stage, intensity, frequency of activation and duration of exposure, confined to certain rock complexes, disturbed geological structures, geomorphological elements and construction sites.

Identification begins with the definition of the lithospheric space of qualitative and quantitative characteristics, typing and mapping of existing manifestations of geological hazards, together with natural and manmade conditions and acting factors (effects) of their formation and development.

As result of identification, the following data are obtained to predict the development of geological hazards:

- areas of development, intensity and repeatability of hazardous geological and engineering-geological processes of certain types in different parts of the estimated area;
- critical characteristics of the geological environment, natural and man-made impacts, in which the formation and activation of hazardous processes, changing their nature, mechanism and intensity in the transition from one stage of development to another;
- areas of development, features of occurrence and physical and mechanical properties of soil massifs, which are the necessary environment for the development of certain types of geological hazards, including those that did not occur within the estimated territory.

In our investigation the following quantitative characteristics of geological hazards are used as the main recorded and predicted intensity indicators:

- areas, amplitudes and speeds of raising and lowering the earth's surface—for the processes of frost heaving, swelling and shrinkage of soils;
- the rate of groundwater level rising and its decline during drainage and pumping—for the process of flooding areas;
- areal rates of annual destruction of the developed territories with different outcomes (removal from land use, local and areal deformations of the earth's surface of a certain amplitude, etc.)—for all types of geological hazards differentiated by

genesis, mechanism, scale of coverage of lithospheric space and (or) other features characterizing their destructive power.

The forecast of development of geological hazards, performed to assess the risks of losses caused by them in various areas and relevant emergencies, is performed by using complex probabilistic-deterministic methods in the following sequence:

- development of verbal models of formation and transformation of geological hazards in certain parts of the assessed area, differing in natural and man-made conditions, existing and possible negative processes, under different scenarios of external natural and man-made impacts with the determination of the probability of their implementation for a given time;
- selection of mathematical deterministic or probabilistic-statistical models (calculation methods) for different stages of development of geological hazards that most adequately reflect their characteristics, and the production of a step-by-step deterministic forecast of events for different scenarios of possible impacts, taking into account the features of the structure and critical characteristics of the geological environment;
- estimation of probability of realization of forecasts of geological dangers on deterministic models (methods) at various combinations of external influences and physical and mechanical properties of soil massifs unstable to these dangers with determination of final results of probabilistic-deterministic forecasting of geological dangers.

Prediction of some little-known geological hazards on pre-project preparation is carried out on the basis of analysis of their manifestations in similar natural and man-made environments using the methods of probabilistic-statistical and expert assessments based on the worst (pessimistic) and the most likely scenario of negative events.

2.3 Risk Calculation

The total differentiated risk of economic losses as result of flooding is recommended to be established, taking into account time of negative impact of this danger on the estimated object by the formula:

$$R_e(S) = P(S)V_e(S)D_e, \tag{1}$$

where $P(S) = T_s/T_c$—the probability of implementation of the process of flooding during the life of the shaft; T_s—the duration of flooding of the facility (years); T_c—service life of the facility (years); $V_e(S)$—economic vulnerability of the object to the flooding process, determined by analogy (year^{-1}); D_e—the cost of the object to defeat the process ($).

The full differentiated risk from slow subsidence and uplift of the earth's surface associated with compaction, swelling, heaving and shrinkage of the surrounding soils is determined by the formula:

$$R_e(H) = P(H)V_e(H)D_cT_c, \tag{2}$$

where $P(H)$—the probability of deformation of a certain amplitude at the end of the service life of an underground object; $V_e(H)$—economic vulnerability of the estimated object for this deformation, determined by analogy; D_c—the cost of the object to defeat the process ($); T_c—service life of the facility (years).

In assessing the risk of these hazards, it is recommended to use, respectively, the average rate of development of the process and the average amplitude of deformations, the probability of which for the service life of the object is taken to be equal to one.

The differentiated social risk from one-time geological hazards $R_s(H)$ is determined in the form of individual and complete values of possible losses of the population with lethal outcome by the following formulas:

$$R_i(H) = P * V_s(H), \tag{3}$$

$$R_s(H) = R_i(H)D_p, \tag{4}$$

where $R_i(H)$—individual risk to die from the danger H, numerically equal to the probability of such event for one person from a group of people within the estimated object (person/person per year); P *—recurrence risk H (cases/year); $V_s(H)$—social vulnerability of the population to danger H; $R_s(H)$—full social risk of dying from the danger H, equal to the number of deaths from this danger during the year (persons/year); D_p—the total population within the estimated object (people).

The geological risk of economic and social losses during the construction of shafts in the mines of Norilsk is generally associated with possible one-time losses as a result of negative and positive deformations of the rock mass due to the sinking technology.

One of the main factors that characterize the conditions of mining operations in the mines the north of Russia is a sharp increase in the disturbance of ores and rocks in the exposed areas of minefields with the rise in the depth of development (Fig. 2). The presence of permafrost, salt layers, aggressive and pressure groundwater, strongly fractured and prone to swelling rocks, significant tectonic disturbance of rock mass can lead to deformation of the lining. Associated with these features, the process of displacement of rocks is complicated by the influence of mining and the strong influence of treatment works carried out within the circumferential safety pillar. During the period of shaft sinking can occur a fall of highly fractured rock in the sites, assigned to a temporary lining, especially in areas with large tectonic faults. In areas of tectonic seams, corrosion of the lining and its destruction may occur [11].

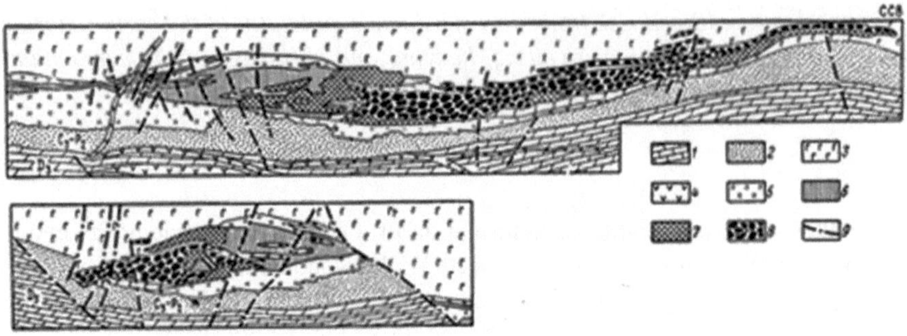

Fig. 2 Geological map. 1—dolomites, marls; 2—sandstones, siltstones, mudstones; 3—gabbro-dolerites; 4—dolerites; 5—contact-modified rocks; 6, 7, 8—ores; 9—tectonic disturbances

The main natural and manmade factors contributing to the destruction of the lining are:

- seismic impact of blasts bore-holes in the bottom of the freshly poured concrete;
- weakening of rocks under the action of water penetrating into the rock through the lining by natural and technological cracks, swelling of rocks in contact with water;
- reducing the strength of the rock mass;
- reduced grip on cracks and sliding of individual blocks on inclined planes;
- change in mechanical properties of the rock as a result of the explosion;
- geological irregularities in thickness of sedimentary rocks, where rocks again changed;
- corrosion of concrete under the influence of aggressive impurities in the water flowing down the shaft, etc.;
- the proximity of the powerful geological disturbances. There is a huge pressure in the fault zone, which breaks the lining. Oil and gas occurrences are also associated with faults;
- mutual adverse influence of shaft and borehole excavations.

Therefore, it is necessary to assess the geodynamic risk. Such an assessment should be based on the results of the analysis of the geodynamic zoning map, which would allow to identify the percentage of accidents on the shaft when they cross the boundaries of geodynamically dangerous zones. The geodynamic risk R_g is determined by the following formulas:

$$R = P_d \cdot P_v \cdot C, \quad (5)$$

where P_d—the probability of occurrence of displacement in this place of the surface and at this time, non-linearly depends on the magnitude of the deformation and is determined on the basis of the method "Hephaestus"; P_v—the probability of vulner-

Table 2 Geodynamic risk

Disturbance zone	Depth, m	P_g
I	141.1 147	0.71 0.94
II	148.1 149.8	0.94 0.81

ability of the object due to the impact of the processes of displacement, non-linearly depends on the magnitude of the deformation; C—maximum amount of damage.

$$R_g = P_g U_g \ (\$/\text{Year}), \tag{6}$$

where P_g—the probability of occurrence of an adverse geodynamic event; U_g—damage from the occurrence of this event.

The probability of an adverse geodynamic event can be estimated by the formula:

$$P_g = \sum_{i=1}^{N} \left(\frac{h_g}{H_i + h_g} + \frac{2L}{\pi (H_i + 2h_g)} \right) \tag{7}$$

where N—number of boundary systems of geodynamic dangerous zones on the computational model; h_g—the width of the boundary of the geodynamic dangerous zone of the i-th system; H_i—the distance between the parallel-oriented boundaries of geodynamic danger zones of the i-th system; L—the length of the shaft or parts of it.

Any anthropogenic impact on such areas without taking into account the state and properties of the geological environment can lead to the activation of these processes and a sharp increase in geological risk at all stages of the existence of the shaft.

Based on the available data, we can determine the probability of an adverse geodynamic event in two areas of the shaft under construction, where tectonic disturbance zones are clearly expressed.

According to the geological section, it is possible to allocate two most weakened zones. Then, assuming that the boundaries of the attenuation zones in width at least extend over the entire diameter of the shaft, in accordance with the formula (7), we have the probability of an adverse geodynamic event (Table 2).

Therefore, the probability of an adverse event for the first zone of tectonic disturbances is in the range of 0.71–0.94, for the second—0.60 to 0.81. In connection with such a high probability of occurrence of a geodynamically dangerous event, it is necessary to conduct research on the development of a reinforced structure of the support, as well as to determine the material and method of conducting plugging works on the disturbed areas.

An important indicator is the “price of risk” in the construction of the shaft, which from a mathematical point of view is the following ratio:

$$S_j = \sum_{t=t_{kc}}^{T} M_t(1+E)^{-t} + P\left[(1-p_d)\sum\nolimits_{t=t_{i1}}^{\Delta t_1} Y_i(1+E)^{-t} + P_d \sum_{t=t_{i2}}^{\Delta t_2} C_{l\pi}(1+E)^{-t}\right] \quad (8)$$

where *Sj*—the amount of monitoring costs(scientific and technical support), the actual damage from the operational risk and the cost of localization of the emergency; *Mt*—monitoring costs (scientific and technical support); *P*—the level of risk, the function of the corresponding component—the share of capital investments provided by the project in the subsystem “investments”; *Pd*—the probability of timely detection of a dangerous situation; *Yt*—the damage on the considered component of risk; *Clt*—costs for localization of emergency situation at timely detection of danger; $\Delta t1$—time of completion of works on elimination of consequences of damage; $\Delta t2$—time of completion of works on prevention or localization of an emergency.

This is what the risk assessment should be aimed at in the construction of shafts. The next step in the environmental risk assessment is to take into account technological and design factors [12, 13].

Hydrogeological conditions of sinking predetermine the flow of minor water inflows into the bottom of the shaft, which increase during the spring-summer snowmelt. It can be assumed that this moisture will contribute to the process of weakening of rocks. The application of sprayed concrete and concrete also contributes to the continuous system weakening of these rocks due to the absorption of water was prepared. It is necessary to determine the time and kinetics of changes in the strength of these rocks and correlate this phenomenon with the time of finding the shaft in the unfixed state of the permanent lining. The lagging of the permanent lining from the bottom of the shaft for leveling geomechanic processes requires time, which can provide saturation of these rocks with moisture and the corresponding weakening, which leads to the risk of falls during the construction of the shaft and reduces the bearing capacity of the shaft lining during the operational period.

In accordance with the geological section of the shaft, a significant part of the most dangerous zone with a depth of 945–2000 m is composed of mudstones, marls, dolomites, clay dolomites. These rocks are characterized by high initial strength (f = 6–12), but have the ability to quickly absorb moisture and reduce the strength characteristics by 50–70%.

In order to be able to calculate the probability of falls from the position of the theory of risks, it is necessary to conduct research on the kinetics of changes in the strength of the rocks when absorbing moisture, as well as the permissible limits of the time of operation of the temporary lining.

3 Monitoring

The main controlled parameters of the state of subsoil in the area of mine placement should be:

- deformation and displacement of the earth's surface or ore bodies (geological layers and rock masses, underground workings, etc.);
- vertical and horizontal displacements of rock massifs accommodating workings;
- the stress state of rock massifs, its variations and related phenomena (rock bumps, emissions, heaving and collapse of the walls of excavations, etc.);
- spatio-temporal variations of fracturing of the medium and its predetermined permeability and fluxes of fluids, gases and heat, hydro-gas-geochemical anomalies;
- seismic phenomena and seismic anisotropy of the medium;
- spatio-temporal variations in the characteristics of geophysical fields (gravitational, magnetic, electrical, acoustic, temperature, radiation).

It is also necessary to constantly monitor potential emergency situations on the basis of the existing legal framework (Fig. 3).

Based on the combination of the main risk factors and their manifestations, the structure of ecological and geodynamic monitoring should include a mandatory set of basic methods.

1. Continuous seismological observations for the study of the seismic regime in the area of developing and following a detailed and micro-seismic zoning. Seismic sounding of geological environment in the areas of abnormal development of seismic and tectonic processes (1–2 times per month, and during abnormal development of seismic and tectonic processes in the mode of daily monitoring). Deformation and displacement parameters are used to generate predictive features of seismic events.
2. Re-accurate leveling to control the current activity of faults, subsidence of the earth's surface and the development of anomalous deformations in the areas of potential focal zones. The results of zonal and local monitoring should be used to develop predictive features of seismic events, anomalous activation of faults and local subsidence.
3. High-precision geophysical observations (gravimetric, geomagnetic), designed to assess the variations of geophysical fields over time and control the deformation and fluid dynamic processes of natural and man-made genesis.
4. Geochemical monitoring together with zonal and local geodetic and geophysical monitoring in areas of potential ecological and geodynamic risk.

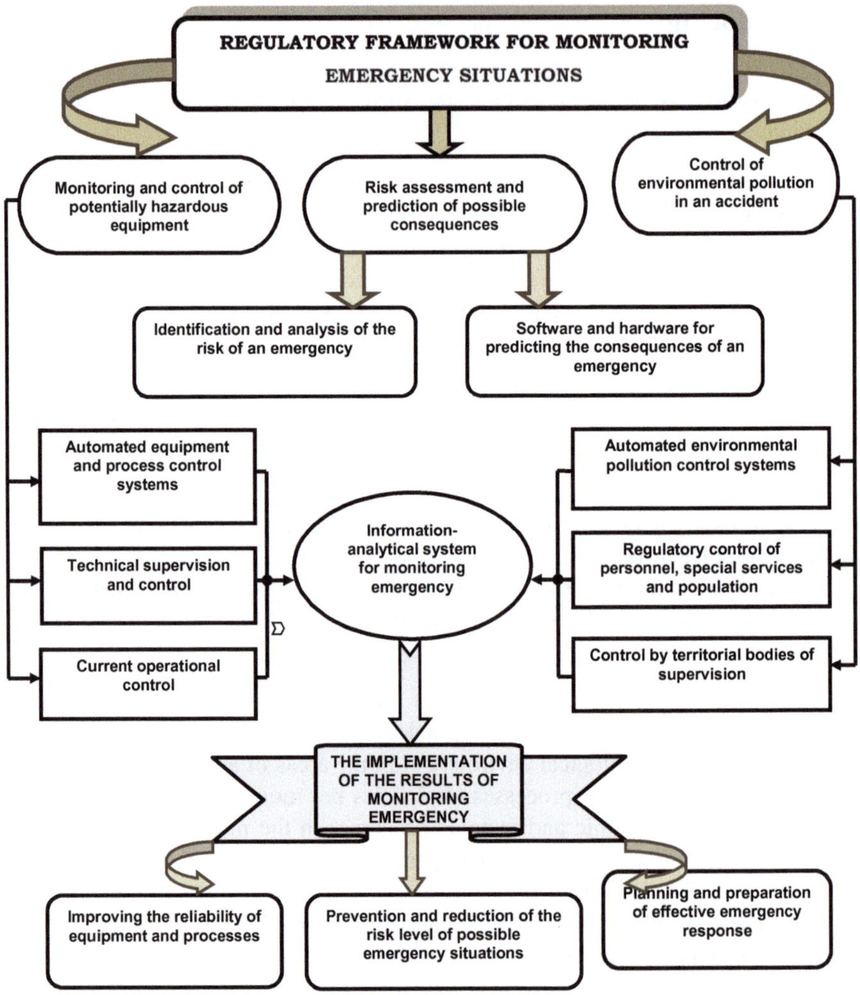

Fig. 3 The structure of the monitoring of potential accidents at the mining enterprises

References

1. Buyanov, V.P., Kirsanov, K.A., Mikhailov, L.M.: Riskology (risk management). In: Tutorial (ed.) Examination, p. 384 (2003)
2. Kisilevskaya, K.E.: Application of the earth remote sensing method for environmental monitoring—the advantages of using remote sensing and geographic information systems for conducting spatial analysis of the earth's surface are described. Min. Inf. Anal. Bull. **2009**(1) (2009)
3. Kolybin, I.V.: Underground structures and pits in urban conditions—the experience of the last decade. Russ. Geotech. **12** (2007)
4. Korolev, V.A.: Monitoring of Geological Environment. Moscow University Press, Moscow, Russian (1995)

5. Kulikova, E.: Environmental Safety in the Development of Underground Space in Large Cities. MSU publishing house, Moscow, Russian (2002)
6. Kolluru, R.: Risk Assessment and Management Handbook, for Environmental. Health and Safety Professionals, vol. 16 (1996)
7. Cohen, B.L.: Catalog of risks extended and updated. Health Phys. **61**(3), 317–335 (1991)
8. Dzuray, E.J., Maranto, A.R.: Assessing the status of risk-based approaches for the prioritization of federal environmental spending. Fed. Facil. Environ. J. **10**(2), 25–42 (1999)
9. Hallenbeck, W.H.: Quantitative Risk Assessment for Environmental and Occupational Health. CRC Press (1993)
10. Young, M., Howard, J.N.: The technical writer's handbook: writing with style and clarity. Appl. Opt. **28**, 4952 (1989)
11. Pleshko, M., Kulikova, E., Nasonov, A.: Assessment of the technical condition of deep mine shafts. In: MATEC Web of Conferences 2018, p. 01021. EDP Sciences
12. Kulikova, E.Y.: Defects of urban underground structure and their prediction. In: IOP Conference Series: Materials Science and Engineering 2018, vol. 1, p. 012108. IOP Publishing
13. Kulikova, E.Y.: Basic directions of improving the durability of the sewage collector lining. In: Applied Mechanics and Materials 2016, pp. 25–32. Trans Tech Publications

Hydro Geopolitics of the Tigris and Euphrates

Nadhir Al-Ansari

Abstract Rivers Euphrates and Tigris are in southwest Asia. The main utilizers of the water of these rivers and tributaries are Turkey, Syria, Iran and Iraq. These rivers rise in Turkey, which makes it the riparian hegemon. Some of the tributaries of the Tigris and Shat Al-Arab Rivers rise in Iran, which makes it the riparian hegemon for these rivers. The lower countries in the catchments are Iraq and Syria and for this reason, they always to ensure the quantity of water required to satisfy their requirements. All these countries are in the Middle East (ME), which characterized by its shortage of water resources. Since the 1970s conflict between riparian counties were noticed due to shortage of available water required, high population growth rate and food security, energy requirements, economic and technological developments and political fragmentation. In addition, there is no public awareness program in all riparian countries and the water management practices are so old leading to high rate of losses. This caused tensions, which sometimes escalated to the verge of war. A mediator is required that is capable to bring all countries concerned to the negotiation table. Syria and Iraq are to give Turkey and Iran some incentives to cooperate. Furthermore, strategic plan based on comprehensive resources development to ensure good water management, minimum water loses, and waste must be adopted by the countries within the basins. This due to the fact that modeling studies of the future suggest that water shortage problem will intensify.

Keywords Hydrology · Shat Al-Arab · River basins risk · Geography

1 Introduction

The amount of water on earth is 1.4 billion km^3 [1] and the annual fresh water required for human being use is about 1000 m^3 annually [2]. About 97% of the available water is saline oceanic water and 77% of the remainder is stored as ice,

N. Al-Ansari (✉)
Water Resources Engineering, Lulea University of Technology, Lulea, Sweden
e-mail: nadhir.alansari@ltu.se

Y. T. Mustafa et al. (eds.), *Recent Researches in Earth and Environmental Sciences*,
Springer Proceedings in Earth and Environmental Sciences,
https://doi.org/10.1007/978-3-030-18641-8_4

and 22% as groundwater and soil moisture, while 0.35% in lakes and marshes. The water within the atmosphere is 0.04%, then there are only 0.01% fresh water supplies in rivers [3] which provide 80% for human beings on the earth and; therefore, rivers carry 0.003% of all the water available on earth [4]. The 80 countries of the Third World that support 40% of the world's population suffer however, from the water shortage problems which has become a daily life fact. These countries suffer from shortage of personal and household needs. Consequently, 1.2 billion people are suffering physically from water shortage and 1.8 billion lack adequate sanitation [4]. Furthermore, in the Third World, about 80% of illnesses and 30% of unnatural deaths are due to water disease and polluted water [4]. Future predictions suggest that there will be 37 countries in 2025 having the shortage of water for all needs [5]. For these reasons, most of the countries of the world try to utilize as much as they can of the water of their rivers to fulfill their demand.

More than 60% of the river basins are shared by more than one state [6]. These basins are located all around the world (57 in Africa, 35 in North and South America, 40 in Asia and 48 in Europe) [4]. They cover 47% of the total land mass on the earth which includes 65% of Asia, 60% of Africa and 60% of South America. Due to the importance of water use and distribution between countries sharing the basins, 300 treaties were signed and more than 3000 treaties include provisions relating to water. Despite these facts, coordinated and integrated management of international river basins is still rare [4]. In the Third World, the situation is the same where more than 165 river basins are shared by many countries [7]. In such basins, there is always a dominant regional power and in the case of Tigris–Euphrates basin, Turkey, is the dominant power [4].

In the ME the average annual rainfall does not exceed 166 mm/year [8–13]. Water allocation per capita does not exceed 500 m^3 in twelve countries [14, 15]. In view of these facts, water resources are very essential to life, socioeconomic development, and political stability in this region. In this work, the conflicting issues on water resources of the Tigris and Euphrates Rivers basins (Fig. 1) discussed and possible solutions to resolve these issues are given.

2 Geography of Tigris and Euphrates Basins

Euphrates and Tigris Rivers originate in southeast Turkey, and some of the Tigris tributaries originate in Iran (Fig. 1). The Euphrates length is about 1178 km in Turkey then it enters Syria and runs about 604 km to reach the Iraqi border where it runs 1160 km toward the south. In Turkey, two tributaries join together 45 km northwest of the city of Elazig to form the Euphrates. The first is Karasu, and the second is Muratsu. The Euphrates River flows through Taurus Mountains and then reaches the Syrian border at Karkamis. Inside Syria three tributaries join the river (Sabor, Belaikh and Khabour). In Iraq, no tributary joins the river. Down south at Qurna city, the left channel of the Euphrates meets the Tigris River forming a new river known as Shatt Al-Arab and the other join Shat Al-Arab further downstream. This river runs

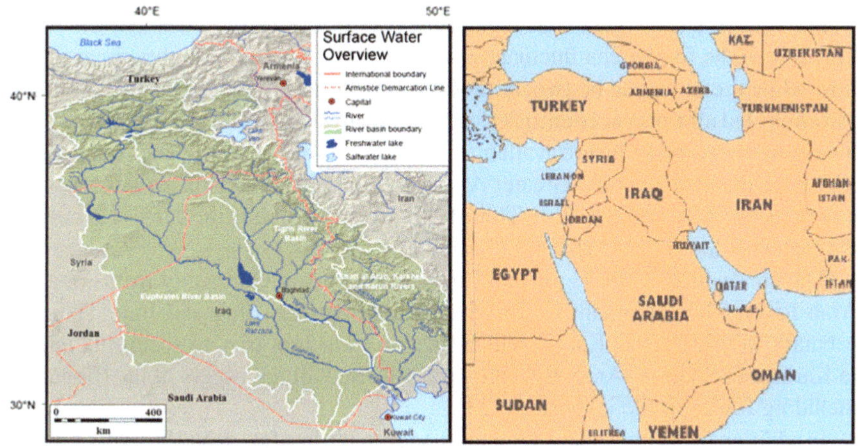

Fig. 1 Tigris and Euphrates rivers basins. Modified from UN-ESCWA [16]

about 132 km until it reaches the gulf. Few tributaries that originate from the Zagros Mountain range in Iran join Shatt Al-Arab. The main tributary is known as Karun, which joins the main river about 32 km downstream Basrah City. The river forms the border between Iraq and Iran for a short distance.

The Euphrates River catchment area in Turkey is about 125,000 km^2 (28.2%), while it is only 7600 km^2 (17.1%) in Syria and the remainder 177,000 km^2 (39.9%) is in Iraq. It should be mentioned, however, that part of the catchment of the river lies in Saudi Arabia (66,000 km^2), but it does not supply any water to the river except when precipitation takes place in that area which is very rare.

High discharges of the Euphrates River take place usually during March–June period where about 63% of the annual flow passes through that period. The minimum flow period is July and August. During flood time, the river carries a huge amount of sediments [17]. Elhance [4] stated that generally the amount of sediments that is carried by the Euphrates in one day can cover an area of 600 acres with a layer 0.25 cm thick. This in fact, how the Mesopotamian plain was formed.

The Tigris River is 1718 km long and its drainage area is 235,000 km^2 distributed between 4 countries as follows: Turkey (17%), Syria (2%), Iran (29%) and Iraq (52%) (Fig. 1). The river rises near lake Hazar in southeast Turkey. It runs toward the Turkish–Syrian border and forms the border between these countries for about 45 km, then it enters Iraq 4 km north FieshKhabour near Zakha city. During its course in Iraq 5 main tributaries join the main river. These are Khabour, Greater Zab, Lesser Zab, Adhaim and Diyala (Fig. 1). Most of these tributaries rise from Zagros mountain range. They carry snow melt and rainfall and for this reason, depending on the phasing of these two types of flows, the flow varies greatly within the Tigris. The sediments transported by the river at Baghdad vary with the discharge from 35 to 52 million ton per year [18, 19].

Flow records show great variation of the flow of the Tigris and Euphrates rivers. The flow of the former can fluctuate where the high flow can be eighty times its low flow while the latter can be twenty times its low flow [4]. The implications of these variations led all riparian countries to build as many dams as they could to overcome these variations and to ensure availability of water (Table 1). One of such the projects, which raised tension, is the Greater Anatolia Project (GAP). This includes 22 dams and 19 hydropower stations [20]. The other reason for building the dams is to mitigate the effect of floods [21, 22].

The confluence of the Tigris and Euphrates Rivers at Qurna city forms Shat Al-Arab River. The catchment of this river shared between Iraq and Iran. The main tributary to this river is Karun, which rises in Iran. Iran tries to secure its interest in Karun and Shatt Al-Arab. Recently, Iran diverted all the waters of the tributaries, including Karun inside its borders [23].

In addition to the above fact, climate change in the last few decades highly affected the region. Droughts affected Syria and Iraq (Fig. 2) [12, 13, 27–29]. Historical flow records for the period 1938–1973 indicate that the Euphrates River mean annual flow was 30 BCM at Jarablus, Syria. After this period, dams were constructed and the flow started to decrease to 25.1 and 22.8 BCM for the periods 1974–1998 and 1990–2010 respectively [16]. Similarly, the Tigris flow at Mosul for the period 1931–1973 was 21.3 BCM and it decreased to 19.1 BCM for the period 1985–2005 [24]. The overall long term flow is decreasing indicating a declining trend of $0.14 \times 109\ m^3$/year for the Tigris and $0.19 \times 109\ m^3$/year for the Euphrates (Figs. 3 and 4) [30].

The effect of dams and construction of hydrological projects in upper riparian countries highly influenced the discharge of the Tigris and Euphrates Rivers (Figs. 5, 6 and 7; Tables 2 and 3). It can be noticed that for the period before the beginning of the construction of dams in the 70s the flow was natural (see Tables 2 and 3). After the construction of dams, the discharges in both rivers decreased (Tables 2 and 3). This can be noticed clearly on the hydrographs of the Tigris at Baghdad where the discharge of the river dropped from 1207 m^3/s (1931–1959) to 522 m^3/s (2000–2013) (Fig. 6). Issa et al. [31] calculated the reduction of flow of both rivers at different stations (Table 4).

In addition, long term records for rainfall prediction records indicate general decrease [33–37]. It is noteworthy to mention that these records (see Fig. 8) indicate that there are years where rainfall will sharply increases in short duration. This will cause intensive flooding. This trend has been noticed on a global scale where there is a significant increasing trend in the annual median of flood durations globally [38].

Table 1 Constructed dams within Tigris, Euphrates basins

Dam	River	Height (m)	Purpose	Completion date
Iran				
Dez	Shatt Al-Arab/Karun	203	I/P	1963
Shahid Abbaspour (Karun 1)	Shatt Al-Arab/Karun	200	P	1976
Masjed Sulaayman (Karun 2)	Shatt Al-Arab/Karun	164	P	1976
Karun 3	Shatt Al-Arab/Karun	205	I/P/F	2002
Karun 4	Shatt Al-Arab/Karun	230	I/P/F	2010
Garan	Tigris/Diyala/Sirwan	62	I	2005
Darayan	Tigris/Diyala/Sirwan	169	I/P	2010
Upper Gotvand	Shatt Al-Arab/Karun	180	P	2012
Lowe Gotvand	Shatt Al-Arab/Karun	22	P	1977
Karkha	Shatt Al-Arab/Karkha	127	I/P	2001
Seimare	Shatt Al-Arab/Karkha	180	P	2013
Khersan 3	Shatt Al-Arab/Karun/Karkha	195	P/F	2015
Turkey				
Çetin dam (Alkumru)	Tigris/Botan	145	P	2016
Aslandağ	Tigris/Greater Zab/Bembo	60	I/M/P (future)	2012
Beyyurdu	Tigris/Greater Zab/Bembo	48	I/M/P (future)	Under construction
Atatürk (Karababa)	Euphrates	169	P	1992
Balli	Tigris/Khabour/Hezil/Ortasu	49	I/M/P	Under construction
Batman	Tigris/Batman	74	I/P	1999
Beyhan I	Euphrates/Murat	97	P	2015
Beyhan II	Euphrates/Murat	62	P	Planned
Birecik	Euphrates	62.5	I/P	2001
Burç Bendi	Euphrates/Gõksu	47	P	2010
Cizre	Tigris/Botan	46	I/P	Planned
Çoukurca	Tigris/Greater Zab/Güzedlere	45.5	W/M	Under construction

(continued)

Table 1 (continued)

Dam	River	Height (m)	Purpose	Completion date
Dumluka	Euphrates/Bugur	30	I	1991
Erkenek	Euphrates/Adiyaman	–	p	Operational
Gŏksu	Euphrates/Gŏksu	52	I	1991
Hecihider	Euphrates/Sehir	42	I	1989
Hancağiz	Euphrates/–	–	I	1988
Ilisu	Tigris	135	I/P/F	2017
Upperkalekŏy	Euphrates/Murat	137.5	P	2017
Lower kalekŏy	Euphrates/Murat	115	P	Planned
Karakaya	Euphrates	158	P	1987
Karkamiș	Euphrates	21.1	P	2000
Kavsaktepe	Tigris/Khabour/Hezil/Ortasu	66	W/M	Under construction
Kayacik	Euphrates/Sajur	45	I/P	2005
Keban	Euphrates	207	P	1974
Kirazlik	Euphrates/Botan	60	I/P	2011
Kralkizi	Tigris/Maden	113	I/P	1997
Musatatepe	Tigris/Khabour/Hezil/Ortasu	34.5	W/M	Under construction
Silope	Tigris/Khabour/Hezil	79.5	W/M/P	2012
Silvan	Tigris/Batman	174.5	I/P	2017
Sirrntiș	Tigris/Birimşe	92	I	2013
Șirnak	Tigris/Khabour/Hezil/Ortasu	56.8	W/M	2012
Uludere	Tigris/Khabour/Hezil/Ortasu	55.5	W/M	Under construction
Syria				
Baath	Euphrates	14	P, I, F	1988
Tabaqa	Euphrates	60	P, I	1975
Tishrine	Euphrates	40	P	1999
Upper Khabour	Khabour		I	1992

Source from Wikipedia [24–26]
F Flood control, *I* Irrigation, *M* Military, *P* Power, *W* Water supply

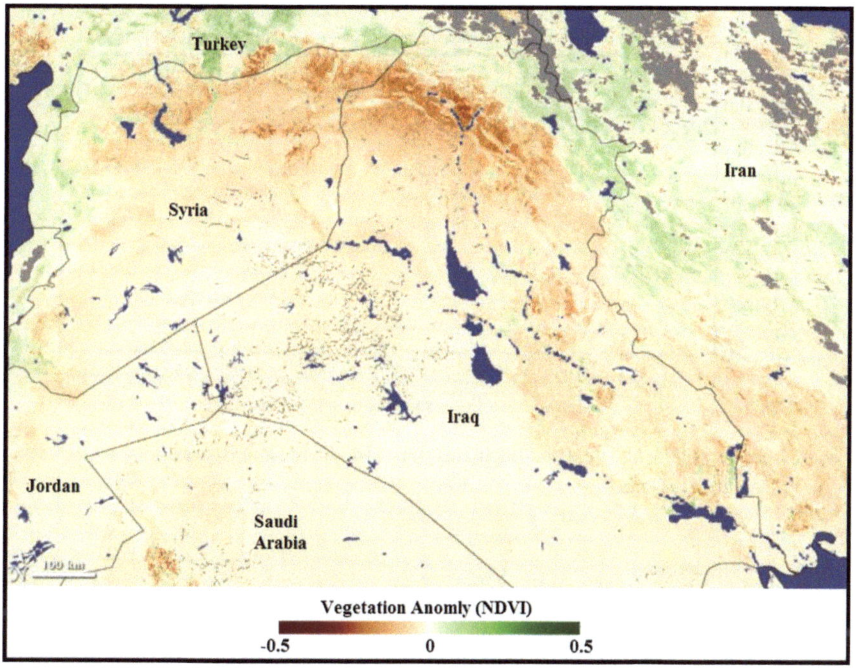

Fig. 2 Drought within Tigris and Euphrates basins. Modified from NASA [27]

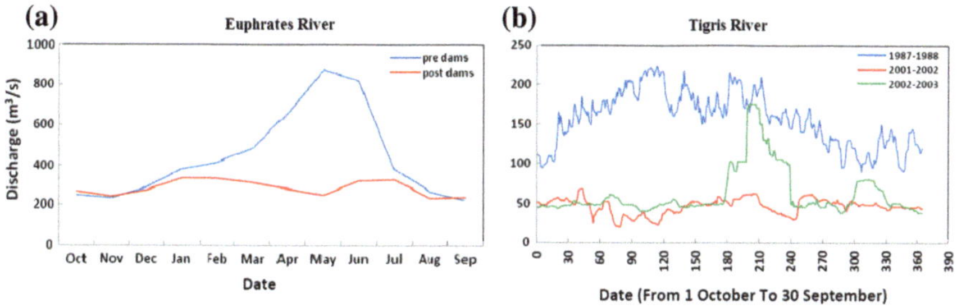

Fig. 3 **a** The Euphrates monthly average flow (Thi-Qar station) pre and post-dams. **b** The average daily flow of the Tigris river near Qurna city before and after the dams construction. Modified after Abdullah [30]

3 Causes of the Conflict

3.1 Water Availability

Different figures are published for the water allocation per capita per year (Tables 5 and 6). The figures given do not include restoring of the marshes in Iraq, and it ignore the situation when GAP project is fully operating. Irrespective of the numbers given,

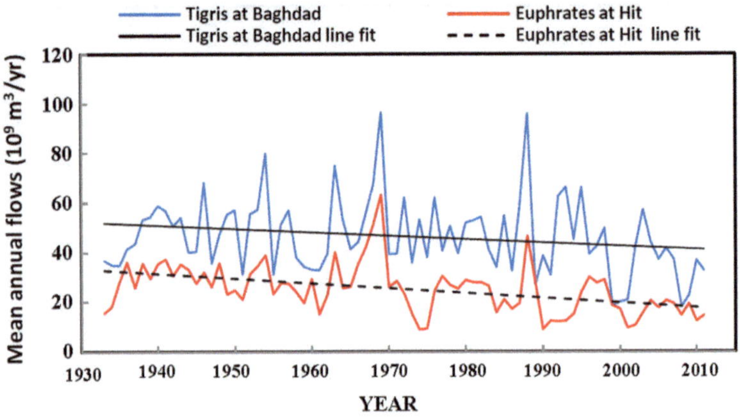

Fig. 4 Long-term flow of Tigris and Euphrates rivers. Modified from Biswas [6]

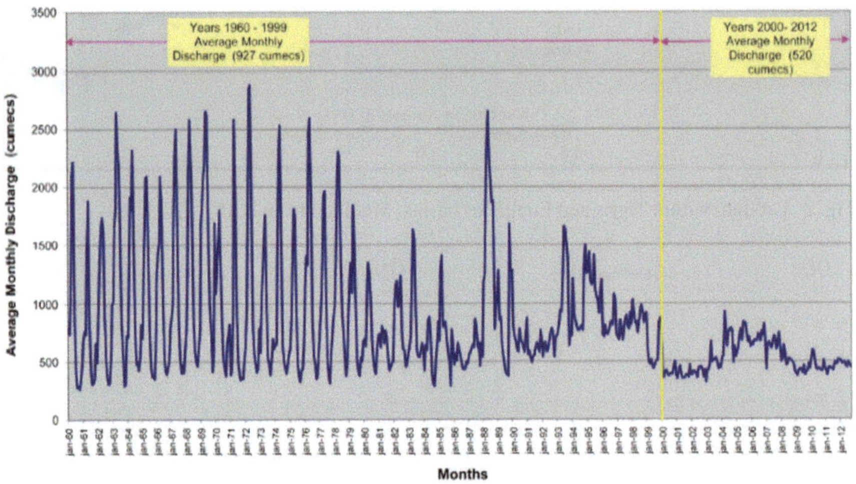

Fig. 5 Water discharge of the river Tigris at Baghdad city for the period 1960–2012. Sources of data until 2007 from Al-Shahrabaly [32]

Turkey claims that the allocations of water per capita in Iraq and Syria are sufficient to fulfill the requested quantities that achieve the people needs in these countries [39].

The flow of the two main rivers is decreasing with time (Figs. 2, 3 and 4). This is mainly due to the Turkish policy of build the dams in the upper parts for the catchment areas of the Tigris and Euphrates and climate change [12, 13, 34, 35, 42–44]. Surface and groundwater resources will be decreased with time [45–47]. Future predictions suggest lower precipitation accompanied with higher temperatures [35]. Serious consequences are expected where (as an example, 71% of the Euphrates River is come from precipitation in Turkey) [48]. This condition will lead to more

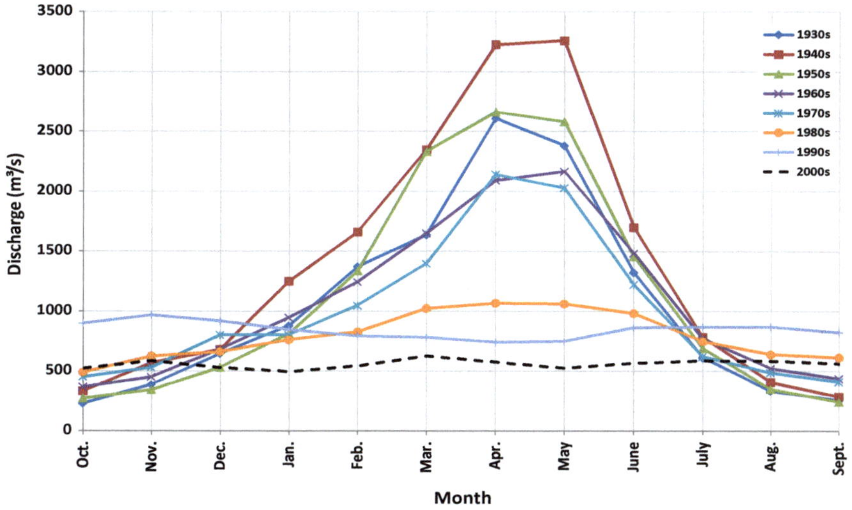

Fig. 6 Decadal hydrographs of the Tigris river at Sarai Baghdad for the period 1930–2013. Some data taken till 2007 from Al-Shahrabaly [32]

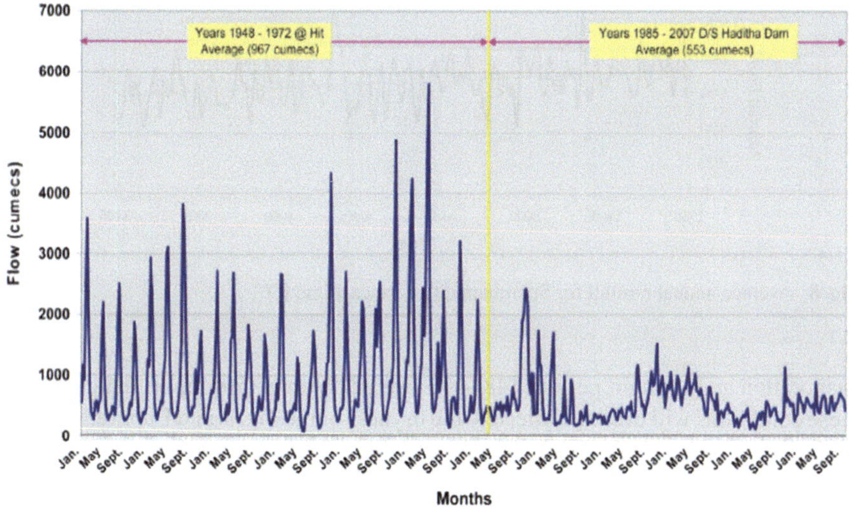

Fig. 7 Water discharge of the river Euphrates at Hit and Haditha cities for the period 1948–2007. Sources of data from Al-Shahrabaly [32]

Table 2 Summary of annual flow volume statistics for the Tigris river in Iraq 1931–2011

Station (drainage area, km^2)	Period	Mean (BCM)	Minimum (BCM)	Maximum (BCM)
Mosul, 56,000	1931–2011	20.0	6.5	43.1
	1931–1973	21.3	11.7	43.1
	1931–1952	19.4	12.2	27.6
	1953–1984	22.0	11.7	43.1
	1974–2005	19.5	6.5	41.7
	1985–2005	19.1	6.5	41.7
Kut, 173,000	1931–2005	25.7	4.2	59.2
	1931–1973	32.0	15.2	59.2
	1931–1952	36.8	15.2	59.2
	1953–1984	24.5	13.2	50.3
	1974–2005	16.7	4.2	47.5
	1985–2005	13.9	4.2	47.5

Modified after UN-ESCWA [16]

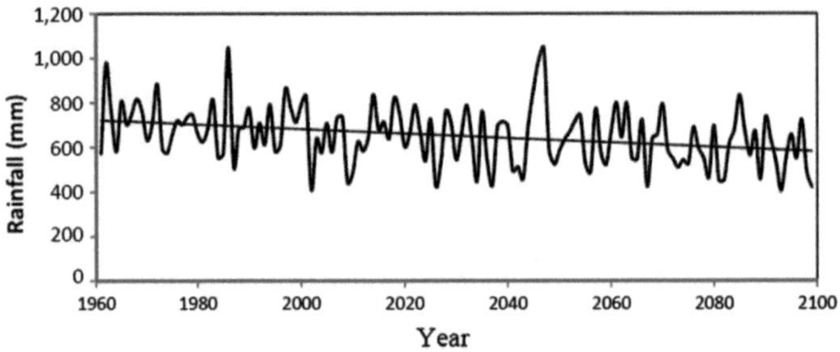

Fig. 8 Average annual rainfall for Sulaimani city northeast Iraq [36]

evaporation and drought periods [48, 49]. United Nations [50] report indicated that these conditions will eventually lead to the dryness of the Tigris and Euphrates Rivers by 2040. In addition, all riparian countries within the catchments of the two rivers will be most water stressed by 2040 [51].

3.2 *Population Growth Rate and Food Security*

Population growth rates are relatively high in the ME particularly Syria and Iraq [52]. Historically, the area (Arab countries only) was populated with about 20 million inhabitants in 1750, and the number in 1996 is 286 million [52]. As far as the four main countries that lie within the Tigris and Euphrates basins (Turkey, Iran, Syria and Iraq) their total population is 221.53 million inhabitants [53–56]. This number will

Table 3 Summary of annual flow volume statistics for the Euphrates river 1930–2011

Station (drainage area, km^2)	Period	Mean (BCM)	Minimum (BCM)	Maximum (BCM)
Jarablus, Syria (120,000)	1938–2010	26.6	12.7	56.8
	1938–1973	30.0	15.05	6.8
	1974–1987	24.9	12.7	34.1
	1988–1998	25.5	14.4	50.1
	1974–1998	25.1	12.7	50.1
	1990–2010	22.8	14.4	32.6
Husaybah, Iraq (221,000)	1981–2011	20.0	8.9	47.6
	1988–1998	22.8	8.9	47.6
	1990–2010	15.5	9.3	20.7
	1974–2010	16.8	8.9	30.7
Hit, Iraq (264,000)	1932–1998	27.1	9.0	63.0
	1938–1973	30.6	15.1	63.0
	1974–1987	23.1	9.3	31.2
	1988–1998	22.4	9.0	46.6
	1974–1998	22.8	9.0	46.6
Hindiya, Iraq (274,100)	1930–1999	17.6	3.1	40.0
	1938–1973	19.8	6.6	40.0
	1974–1987	15.3	3.1	24.1
	1988–1998	13.8	7.7	27.9
	1974–1998	14.7	3.1	27.9

Modified after UN-ESCWA [16]

Table 4 Annual reductions in water inflow of main gaging stations

Station	Average annually inflow		Average annual water reduction		Annual percentage reduction %
	$m^3\ s^{-1}$	km^3	$m^3\ s^{-1}$	km^3	
Mosul, TS_3	569.75	17.96	1.35	0.0426	0.2372
Beiji, TS_6	1295.94	40.87	8.64	0.272	0.666
Baghdad, TS_8	979.26	30.882	9.73	0.307	0.994
Kut, TS_{11}	815.50	25.72	14.73	0.464	1.804
Husaybah, ES_1	708.30	22.33	1.57	0.0495	0.222
Hit, ES_2	802	25.292	7.72	0.243	0.961
Hindiya, ES_3	551.62	17.40	4.02	0.127	0.730
Nasiriya, ES_4	430	13.5	1.452	0.0458	0.339

Modified after Issa et al. [31]
TS gauging stations on the Tigris river, *ES* gauging stations on the Euphrates river

Table 5 Water allocation per capita per year in Turkey, Syria and Iraq

Country	Water allocation (m^3/capita/year)			
	1990	2000	2010	2020
Turkey	3223	2703	2326	2002[a], 980[b]
Syria	1636	117	880	760[a], 780[b]
Iraq	2352	1848	1435	1062[a], 950[b]

Source of data [a]Bilen [40]; [b]Affairs [41]

Table 6 Population characteristics within Tigris–Euphrates basins

Country	Population (million)	Rate of growth (%)	Projected population (million)		Percent urban
			2025	2050	
Turkey	81.91	1.45	86.12	95.62 95.819[a]	71
Syria	18.28	3.7	23.41	34.02 34.90[a]	75
Iraq	39.33	2.78	47.19	81.49 83.65[a]	66.9
Iran	82.01	1.05	86.72	93.55 92.21[a]	73.8
Total	221.53		243.44	304.68	

Modified from Drake [52] and Worldmeters [53–56]
[a]Wikipedia, the anticipated population growth for several Middle Eastern countries (United Nations, medium fertility variant)
https://en.wikipedia.org/wiki/List_of_countries_by_future_population_(United_Nations,_medium_fertility_variant)

increase by about 10% in 2025 and about 37% in 2050 (Table 5). Accordingly, the allocation of water per capita will decrease too [57]. The allocation within the Tigris and Euphrates basins is about 975.3 m^3/year/capita now (Table 6) and this will drop to 887.6 and 709.2 m^3/year/capita in 2025 and 2050 respectively. It is noteworthy to mention that there are other references that give different figures, but the outcome is the same where there will be decrease in water allocations with time.

Governments in the ME try to attain food self-sufficiency and for this reason, the agriculture is considered as the largest consumer for water, it is consume 66% of the total demand [10, 58]. Countries within the Tigris and Euphrates basin allocate as an average 84.3% of the water consumption for agricultural purposes (Table 7). Iran has the maximum water allocation (92%) while Turkey (73%) has relatively the minimum water allocation for agriculture [67–70]. Thorough consideration of agriculture is required to objectively analyze and adequately address the water shortage problem [59]. However, this is not the case where countries have extremely ambitious goals to secure the self-sufficiency of food, and they require a core changes in water management policies and widens their national outlook [60]. Recently, achieving the increasing demands for water represent a severe challenge because it is over the

Table 7 Water use in countries within Tigris and Euphrates basins according [67–73]

Country	Water allocation per inhabitant (m^3/year)	Cultivated area (ha)	Water withdrawal (10^6 m^3/y)			
			Total	Irrigation + livestock	Municipalities	Industry
Turkey	563	26,606,000	40,100	29,600 73%	6200	4300
Syria	921	5,742,000	16,690	14,669 87%	1426	595
Iran	1356	18,107,000	93,300	86,000 92%	6200	1100
Iraq	2632	6,010,000	66,000	52,000 79%	4300	9700
Total			216,090	182,269 84.3%	18,126 8.4%	15,695 7.3%

abilities of each individual country [59]. Turkey is trying to convert the area of the GAP into a breadbasket, and this threatens the irrigation based agricultural potential of the lower riparian Syria and Iraq [61]. Iraq and Syria were exporting the grains for different countries, but now they are importing their needs for grains, and their agricultural production decreased to become less than Turkey [61]. Syria tried long time ago to achieve food self-sufficiency and to increase its irrigational areas; drip irrigation was used. As a result, severe reduction in wheat yield has been occurred; it was a decrease up to 50%. In addition, huge portion of livestock died due to water scarcity. Consequently, many individuals joined to the insurgents to save themselves and their families [62]. Iraq tried to increase its agricultural lands and become a grain exporter again by 2017 [63, 64]. Numerous projects were executed but salinity and water logging created serious problems for agricultural activities. After the second Gulf war, Iraq is importing its food reflecting disastrous agricultural conditions [65, 66]. Recently, food security and self-sufficiency are not a major concern in both Iraq and Syria where national security problems, especially the threat of ISIS is the priority now.

3.3 Energy Requirements

Iraq is oil exporting country since the beginning of the twentieth century, while Syria started to export oil in 2001 and, Turkey has no oil reserves [74]. For this reason, Turkey is trying to reduce its dependence on oil imports as an energy source. To achieve this goal, Turkey is trying to use hydroelectric power to cover as much as 40% of the required energy [75]. The GAP project is one of the strategies used so that Turkey can reduce 28 million ton of its oil imports when this project is fully operational [76].

Syria and partially Iraq relies on hydropower to generate electricity. Despite the fact that Syria is oil producer but it relies on hydropower to generate electricity. This fact gives the opportunity for Turkey to decrease the water quantities which release from the Euphrates through the GAP project and put Syria under threaten. Although, the Turkish Government declared several times that the GAP is purely a development project. Some people believe there are several external and internal strategies involved within the implementation of the GAP project [77–81].

3.4 Water Management

Poor water management strategies have exacerbated the water scarcity problems within riparian countries [57]. Water is wasted through old irrigation techniques where flood irrigation is still the dominant method used. In addition, the irrigation canals are unlined and/or uncovered, which enhance water losses. Water qualities of the rivers are deteriorating due to the extensive use of chemical fertilizers and

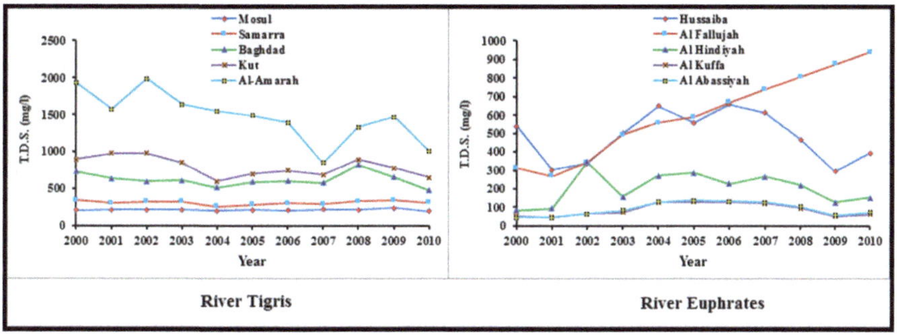

Fig. 9 Salinity variation along the Tigris and Euphrates rivers for the period 2000–2010. Sources of data from Bureau [82]

pesticide. Back flow from irrigated areas and dispose of the industrial and municipal wastes into the rivers is accelerating the contamination of these rivers. Recent data for the period 2000–2010 collected by the Consulting Engineering Bureau at Baghdad University [82] show the steady increase of the salinity in both rivers (Fig. 9). Similar trend was noticed by UN-ESCWA [16] for long term salinity trend in both rivers. TDS is about 300 ppm at Ataturk Dam on the Euphrates, and it increases to 600 ppm at the Syrian–Iraqi borders which are much more than the recommended TSD concentration for irrigation, and it is continued to increase for more than 1200 ppm (minimum) downstream in the Iraqi southern. Specifically, at Samawah [83, 84]. As far as the salinity within the Tigris River, it increases dramatically downstream Baghdad due to intensive irrigation. To overcome the salinity problem in Iraq, a main outfall drain (MOD) was constructed south of Baghdad to the Gulf for a distance of 565 km to carry drainage water from irrigation projects from 150,000 km^2 with a discharge capacity of 210 m^3/s [85]. To overcome salinization and water logging, huge networks and sub-surface tile drains, and surface drainage canals were constructed to collect drainage water from agricultural fields to be dumped in MOD [86]. Taking all these measures, recent estimates indicate that 4% of irrigated areas are severely saline, 50% are of medium salinity, and 20% are slightly saline [87]. Salinity increase in conjunction with decrease of flow downstream along the two rivers stream has been adverse effects on the agricultural areas which they located in the south of Iraq. This situation generates a resentment and frustration and lead to raise the irritation that might cause a conflict.

3.5 *Economic Development*

The ME is going through a development stage which caused the movement of about 50% of the population from rural to urban areas. Such movement aggravates the issue of water shortage where water consumption increased about 10–12 times its normal per capita as village dwellers [52]. Furthermore, the relative fast increase in oil prices caused rapid economic developments and raised the standard of living in Iraq and Syria [57], although the economies of both countries are hardly affected by corruption and the struggle with ISIS in the past few years. The two countries raised their need for water in view of these developments. When the claimed needs for Turkey, Syria and Iraq are added it sum up to 149% of the total water available [74]. Since Turkey is not considered as one of the countries that produce oil, it is trying to use its water as a commodity for bargaining where in 1992 the Turkish president was announced the ceremony of opening the Ataturk dam that "Neither Syria nor Iraq can lay claim to Turkey's rivers any more than Ankara could claim their oil. The water resources are Turkey's; the oil resources are theirs. We don't say we share their oil resources, and they can't say they share our water resources" [88]. Furthermore, Turkey proposed Peace Pipeline and Manavgat River project focus to trade water with Mediterranean and ME neighbors [89].

3.6 Technological Development

The riparian countries built several dams on the Tigris and Euphrates Rivers and are planning to build more dams. The ME well known with its high temperatures and the construction of these dams have increased the quantity of evaporation from the surface water of the reservoirs. Furthermore, agricultural practices still not modernized where old irrigation methods are still used. Such practices are also leading to high quantities of water losses. Syria exerted efforts to use modern techniques in irrigation systems and it faced plenty of problems [90, 91]. One of the main problems that farmers were not educated and could not understand and apply the new technologies.

3.7 Political Fragmentation

ME was dominated by the rule of Ottman Empire since the thirteenth century till its defeat and dissolution during World War I. Then the area was divided into different countries, but Britain and France ruled them. During these periods, less conflict took place among the people of this area. Afterwards, the ethnic tendency for the region increased, which contributed in raising the disparities between these country's people. Consequently, the political and economic competition became more sever and the people more nationalistic. The tension and friction between the United States of America and the Soviet Union and their allies during what is known as the "Cold War" had a restraining impact on the chance of major conflicts, and this does not exist now.

Excessive use of water resources and water pollution became main factors for tension and clash. The 1967 war between Arab states and Israel is an example that reflects this fact where water was one of the hidden reasons. As well as the Israeli occupation of Lebanon in 1982 where they controlled Litany River and diverted its stream. To meet the high water demand, Israel is extracting 40% of its water rom aquifers beneath the West Bank and Gaza [52]. Plenty of dams were built on the Tigris and Euphrates Rivers and their tributaries (Table 1). Unilateral decisions without any consultation with riparian countries also raised friction [30]. As an example the tension between Syria and Iraq in 1974 over the Euphrates water sharing. Future prediction models for surface water and groundwater resources show their depleting in the ME [11, 16, 45–47, 92–94]. For these reasons, UN Secretary General Boutros Boutros-Ghali said in 1985 that the next war in the Near East would not be about politics, but over water [95].

3.8 International Water Laws

The International Law Commission of the UN worked on the Convention on the Law of the Non-Navigational Uses of International Watercourses for three decades, and it was approved by the United Nations General Assembly on the 8th July, 1997. Three countries voted against this law. These countries were: Turkey, China and Burundi [4]. This convention needs however to be ratified by thirty-five countries in order to enter into force, which it had not attain hitherto. In this law, the UN stated rights and obligations that states could follow. It also gave the principals and mechanisms that states should follow to avoid dispute escalating to the level of acute conflicts. This law could be adequate for non-arid zones and not for an arid region such as ME [13]. In addition, despite the principals stated, there are no international legal commitments to force the countries to share their water [96]. Having this situation, then agreements will depend upon several factors like: the goodwill of the countries which they shared the drainage basin, the available internal and external power and the national benefits for the country to keep going to its politics [97].

3.9 Public Awareness

Despite the fact that Syria and Iraq are facing water shortage problems now, it is expected that all countries within Tigris and Euphrates basin will experience the same situation in the future. This is due to increase of population and development in these countries. This implies required improving the present water supply efficiency, and demands to fulfill the sustainability through secure the required water for future generations. To achieve such goal, all parties concerned are to be involved [98].

A strategy to be adopted to construct a comprehensive public awareness program about water, which comprises promotional and practical activities, and observing and assessing their effectiveness. Al-Ansari [13] suggested educating the decision makers in the water sector such as water planners, managers and marketers; and the politicians who involve in set the external and internal water policies; and the

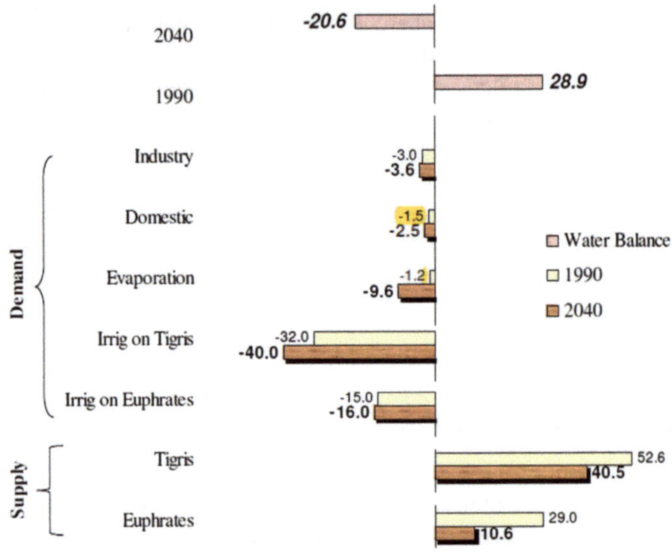

Fig. 10 Iraqi water balance 1990, 2040 (BCM) [99]

educators about the significance of water conservation in the sector of potable water supply and how it may be approached. Subsequently, they will be contributed in transfer the benefits of the awareness program into society individuals as a whole. Special syllabus in schools is to be designed to increase awareness for water significance by developing and finding methods to present this subject, and the media should have a vital role in identifying the importance of water issues. Farmers are to be trained on utilize of modern irrigation systems which they are convenient for arid regions since the agricultural sector is the highest consumer of water resources. Using non-conventional water resources should be taken seriously. The public should understand the importance of proper water management. The ignorance of the impact of political and economic decisions of the long term guarantee of water resources is one of the biggest problems in the ME [13].

Texas Water Development Board (2010) set a program for water conservation and use that can be adopted by the countries concerned. The main points in this program can be summarized as follows:

- What are the legislation that applied in granting the permits for water and wastewater?
- How is the water produced and distributed?
- What is the methods of collecting and treating the wastewater?
- What are the services' quality and methods of maintenance, which are provided by your utility?
- What kind of customer service does your utility provide?
- What conservation measures are in place?

As an example, it is noteworthy to mention that Iraq is expected to have −20.6 billion m^3 in 2040 (Fig. 10) [99]. Such a figure raises the alarm, and action should be taken starting now.

4 International Agreements

4.1 Historical Background

Turkey, Iran, Syria and Iraq are the main riparian countries within the Tigris and Euphrates basins. Since Syria and Iraq are the downstream countries within the basin, so they are always trying to ensure the required amount of water that can meet their domestic, agricultural and industrial demands. In addition, they consider that these basins as international "watercourses" which should be treated as an integrated entity by all the riparian users. On the other hand, Turkey considers the Euphrates and Tigris Rivers as "trans-boundary Rivers" where both are under Turkish authority until they across the border, and when they united together to form Shatt Al-Arab River then it becomes an international river. In addition, in 1997, three countries voted against the International Law Commission of the United Nation on the law of Non-Navigational Uses of International Watercourses one of them is Turkey. Turkey also considers this is not legally binding because the convention does not apply to them [100]. Furthermore, Tigris and Euphrates Rivers basins are considered as one basin by Turkey according to the fact of formation Shatt Al-Arab River from their gathering, in addition to mass water transfer between the Tigris and Euphrates Rivers via the Tharthar system, while Syria and Iraq suppose that each river had independent basins.

Historically, the Tigris and Euphrates basins were under unitary authority of different empires and colonies [4]. Water issues disputes that took nationalistic character started after the British and French mandates were dissolved. Before World War II, the first signed treaty was in 1913 between Britain, Russia, Iran and Turkey for the regulation of Shat Al-Arab River. France and Great Britain as a represented power for both of Syria and Iraq, respectively, they signed an agreement in 1920 to establish a committee that coordinates the efforts toward the utilizations of the Euphrates and Tigris Rivers [16, 101]. This was followed by two treaties in 1921 where it states in article 12 that Aleppo city can use the water of the Euphrates River. Then in 1923, another treaty was signed between Allied powers and Turkey known as Lausanne agreement concerning the Euphrates and Kuviek Rivers. The treaty also included a provision that Turkey must consult Iraq before undertaking any hydraulic works (article 109) [16, 101]. In 1926, Turkey and Allied powers signed a Neighborly Relations treaty where they agreed to cooperate together to use the Euphrates basin. The commission on the demarcation of the Turko–Syrian Frontier on the Tigris was established in 1930. As a result, a treaty took place between France as a dominant power on Syria from a side and Turkey from another side, which stated that the border between the two countries follow the thalweg principle, establishing the border in the middle of the Tigris, regardless of the changes in the river's stream [16].

Iraq declared its independence from Great Britain in 1932. Then, Iraq and Iran placed an agreement in 1937. This was for demarcating their border and regulating navigation in Shat Al-Arab River. Now all the aforementioned treaties have much importance for contemporary interstate relations and geopolitics in the basins [4, 16]. In the 1946, Ankara Treaty of Friendship and Good Neighborliness was signed by Turkey and Iraq [102] and considered as the first bilateral cooperation between both countries regards management the common water resources. The treaty is stipulated that Turkey has the right to install and operate permanent flow measurement facilities and any obtained data should provide to Iraqi side (article 3), as well as inform Iraq with any intention of construct any water projects [16, 101]. Turkey promised that it would not alter the Euphrates flow without informing Iraq, and, to adapt any future works to the needs of both states. In that treaty, Iraq was allowed to construct protection and observation posts in Turkey's territory to prevent downriver flooding [103].

First few hydrological projects began in Iraq in the 1950s where Samarra barrage and Dukan Derbendikhan dams were constructed in Iraq [101, 104]. The first meeting between Turkey, Syria and Iraq took place in 1965 where it was decided to demise and end of the treaty system [105]. New phase of their relationship took place between the three riparian countries in 1960s when Turkey decided to construct Keban Dam. Turkish and Iraqi experts held a meeting in June, 1964 and in that meeting, Turkey agreed to retain the discharge at the downstream of the dam at a flow average 350 m^3/s, the natural flow of the river can provide the adequate quantities to maintain the discharge at this average. Turkey proposed establishment of a Joint Technical Committee (JTC). This committee duty is to investigate the rivers to estimate the average of annual discharge for each river and to determine the required water quantities for irrigation for beneficiary countries through joint field studies. The main procedures and outlines of the committee works should be documented in order to facilitate an agreement on water rights [106].

In 1965, a tripartite meeting was held in Baghdad. During that meeting, Iraq, Syria and Turkey demanded 18, 13, 14 BCM of the Euphrates water annually. This amount exceeds the low annual flow of the Euphrates River. In these meetings, proposed dams were discussed, in particular, Keban (Turkey) and Tabaqa (Syria) Dams. After 22 rounds of talks, it came to a standstill [107]. One of the main issues that were proposed by Turkey is that it agrees to sign a tripartite treaty only if there was an "inclusive agreement on the distribution of the waters of all the rivers common to it and Syria" [108]. Later, Keban Dam was operating in 1973 and Tabaqa Dam in 1974. This raised very high tension between Iraq and Syria. The latter promised Iraq for a supply of 200 MCM from Tabaqa Dam. The tension between Iraq and Syria became very high again in 1975 when Syria started to impound Assad Lake, and Syria denied the Iraqi right in sharing a part of the water of this lake where Syrian side declared that the coming flow from Turkey was less than the normal flow, and they received half the expected quantities of water [101]. The Arab League, Saudi Arabia and Egypt tried to mediate and solve the problem, but all their efforts failed and both countries amassed troops along their border in June, 1975. Later, Saudi Arabia suggested sharing the water of the Euphrates between both countries based on the water quantities that received by Syrian from Turkey; this proposal reduced the dispute, but they did not sign any agreement. Turkey expanded the Lower Euphrates Project and its name became Güneydogu Anadolu Projesi (GAP), the Southeastern Anatolia Development Project (GAP) in 1977 [104]

In 1980, both Iraq and Turkey signed a cooperation protocol in technical and economic levels, later on Syria joined to this protocol. Specifically, in 1983. Issues of regional waters—particularly the Euphrates and Tigris Rivers—were to be discussed by a joint technical committee (JTC). Later, Syria supported the insurgent from the Kurdish and Armenian nationality against the Turkish regime and allowed them to attack GAP projects to have their headquarter in Syria. These rebels were conducting subversive actions on the GAP projects' works [109] while, Iraq gave permission to the Turkish troops to carry out their attacks against the Kurdistan Worker's Party (PKK) on Iraqi lands. In 1987, Syria and Turkey signed a protocol for Economic Cooperation. In that protocol, Article 6 reads as follows: "During the filling up period of the Atatürk Dam reservoir and until the final allocation of the waters of Euphrates among the three riparian countries, the Turkish side undertakes to release a yearly average of more than 500 m^3/s at the Turkish–Syrian borders and in cases when monthly flow falls below the level of 500 m^3/s, the Turkish side agrees to make up the difference during the following month". In addition, Article 7 of the protocol states that Turkish and Syrian sides shall coordinate their cooperation with Iraq to achieve fair allocation for Euphrates and Tigris water within the less potential time. Article 9 confirms the determination of the two countries to construct a hydropower project and operate the irrigation systems together on both rivers [16, 109]. Then Syria and Iraq agreed that 58% of the total water quantities that received by Syrian from Euphrates River would be released to Iraq [110]. Due to the fact that Iraq was ignored and was not asked to sign that protocol, it did not allow Turkey to attack the PKK in Iraq and in 1988, Iraq suppressed its Kurdish uprising in February 1988, 60,000 Kurds fled to Turkey, further deteriorating their relationship [102].

Turkey notified its downstream neighbors before November 1989 that it is going to impound Ataturk Dam's reservoir. It explained the technical reasons behind the action and also provided a detailed program for the replenishment of the losses. In addition, delegations were sent to the region to explain the need for the action, and the measures taken. Impounding started on the 13th January, 1990 and ended to February 13, 1990. January was chosen because the demand for water is low in that month. Official complaints against this action were registered by Iraq and Syria and called for a new agreement to share the waters of the Euphrates River. They also agreed that 58% of the Euphrates water that Syria receives would be provided to Iraq [104]. Iraq and Syria protested against the construction of Birecik Dam in Turkey, which raised friction and tension again in 1996 [104]. In view of this situation, Iraq and Syria organized a joint water coordination committee to face water shortage problems. They agreed that Tigris and Euphrates water can be used in an equitable, reasonable sharing and utilization. To resolve the conflict, Turkey asked Syria in May, 1996 to engage in talks, and it suggested that water can be divided according to the area of cultivated lands while Syria asked for equal share [111], consequently, no agreement was reached. Syria continued to support the Kurdish rebel group (the PKK) to attack southeastern Turkey from Syrian soil. Turkey retaliated to this 1987 act in October, 1998 and asked Syria to stop supporting terrorists immediately, which was understood as a threat of military intervention. Syria responded to Turkey, and they signed what is known as the Adana Accord. Accordingly, the relationship between the two countries improved, and they signed another agreement in 2001 between GAP's Regional Development Administration (GAP RDA) and General organization for Land Development (GOLD) [101].

In 2002, Syria and Iraq signed an agreement which allows the former to establish a pumping station on the Tigris River. Project area and volume of water extracted was specified in that agreement [16]. Later in 2007, Turkey and Syria reactivated the JTC and held a series of meetings during which they agreed to share information on meteorological patterns and water quality. The amount of water that Turkey released to Syria and Iraq was effected by the drought that was experienced in the region through the period 2007–2009. In 2009, Turkey and Syria signed a new agreement known as "Strategic Cooperation Council Agreement" and number of MoUs were signed. All the signed agreements focused on emphasis on improvements to water quality, the construction of water pumping stations (on the Syrian stretch of the Tigris) and joint dams, as well as the development of joint water policies [16]. Turkey assisted Iraq through that period with additional water, but they did not sign any agreement [97].

Some of the tributaries of the Tigris rises in Iran and as far as Shat Al-Arab, is concerned, the two main riparian countries are Iraq and Iran. In this context, the first agreement was signed in 1913, which is known as the Constantinople accord concerning borders delineation between the two countries which dealt with continuous shifting of the river watercourse. This was followed by another agreement in 1937 signed with support from the League of Nations. According to these agreements, Iraq has the full sovereignty over the two banks of Shat Al-Arab River, although Iran kept claiming half of the river to its sovereignty. Iran supported the Kurdish rebels in north of Iraq during the 1970s so that it can exert pressure on Iraq to negotiate the Shatt Al-Arab status. In 1975, an agreement was signed in Algeria between the two countries. As a consequence, Iraq to concede its right in half the Shatt Al-Arab and the Kurdish rebellion was ended. Iraq felt that it was humiliated, and continuous strain started to increase till first Gulf war started between Iraq and Iran in 1980. The war ended without resolving the problem of Shat Al-Arab [112]. Till now, all Iraqi governments kept the strong position that Iraq would never reinstate the 1975 agreement.

The improved relationship between the two countries after 2003, had led the Foreign Ministers of the two countries to meet and discussed among other things the Shatt Al-Arab issue in 2014. Both parties agreed to move forward and work on the marking of land borders and to implement agreements in accordance with the borders' treaty, protocols and agreements that were signed between the two countries in 1975 [113]. This is very important because Iran has done several things that affected the water resources' situation in Iraq [30] which can be summarized as follows:

- Dam was built on Wand River in 1960, and as a result Khanaqeen city was cut from its water resource. Three more diverting dams were built on the same river later.
- Diverting Serwan River waters which is one of the tributaries of Diyala River.
- Dams were built on valleys near the border with Iraq to divert the water inside Iran.
- Building dams on Karkha River.
- Water projects on Karun River and diverting its water inside Iran.

These acts caused considerable hardship to the Iraqi population in general and to the inhabitants of border areas, in particular; such as Khanaqeen, Mandli, Badra and Jassan and lately to Qala-Diza, Halabja, and Shir-Zur in Iraqi Kurdistan Region.

4.2 International Law and Water Sharing Issues

Disputes concerning water resources of the Tigris and Euphrates Rivers between riparian countries seriously started in the 1970s when some of the riparian countries started to build dams on these rivers and the effect of droughts that dominated the region. Each country has its own justifications and explanations to the laws concerned, which are the International Water Law, the Helsinki Rules, Berlin Rules [114] and the UN convention on the law of the None Navigational uses of International Water Courses [115]. These laws are based on restricting the territorial sovereignty of any riparian state to the part of an international fresh water system that is located on its territory, and the riparian State has to respect the right of the other riparian states to utilize the system.

In addition, they approach the problem through the theory of community interest, and the theory of limited sovereignty to reflect the interdependent character of freshwater systems. In view of these theories, two provisions are to be considered. These are the doctrines of equitable utilization implying fairness and reasonable use and the rule of causing no harm. Therefore, riparian states should recognize the limitation imposed by the hydrological cycle (i.e. planned amount of water withdrawal from a freshwater system does not exceed the amount it receives through the hydrological cycle), and the water should be capable of regeneration to the hydrological cycle (i.e. not polluted) [116].

Syria and Iraq claim that Turkey is having water more than it needs, and they would like to have more water to secure the water demand for their uses. Turkey claims that this is not correct because out of 180 BCM of the annual runoff, only 110 BCM of water is usable and 25.9 BCM can be made available. These figures are based on considering technological, topographical and geological reasons and this makes this resource unavailable sometimes [117–119]. When allocation of water per capita is considered (Table 7), it should be noted that these figures will change due to population growth rates (Table 6) and effect of climate change.

Historic rights for the use of water of the River Euphrates were claimed by Syria and Iraq. The Turkish response was rejecting these claims where according to Helsinki rules; Articles I and V [114] acquired rights can only be considered if it is based on equitable use for socioeconomic, geopolitical and hydrological factors, in addition to the avoidance of unnecessary waste in the utilization of waters of the basin. Furthermore, Turkey always accuses Syria and Iraq for using old irrigation techniques and wasteful water management procedures. Turkey suggested a three stage plan for the technical talks of the three riparian countries to solve the water allocation problems [120].

Allocations and determining the demands for each country are to be based on: (1) assessment of available water resources (2) conducting inventory studies of available land resources and (3) improving irrigation practices to determine their economic viability. This plan was based on two premises. These are to consider the Tigris and Euphrates Rivers as one transboundary water course and secondly is that water requirements are to be based on scientific studies of the needs of that country. Iraq and Syria rejected this plan where they considered it as being vague and gives advantages to Turkey and fringes on the sovereignty of the riparian states. In addition, Turkey considers the two rivers form one basin and deficiency of water in the Euphrates can be avoided by transferring water from the Tigris to the Euphrates via Tharthar canal. Iraq rejects this idea and considers the two rivers in geographically separate basins. In any case, this idea of a single basin is not valid any more after the construction of Ilisu Dam and the effect of drought due to climate change, which has been experienced recently.

Syria and Iraq accuse Turkey of ignoring the "causing no harm" doctrine defined in article X of the Helsinki Rules, and article 16 of the Berlin Rules [114] in addition to article 7 of Part II of the UN convention on the law of the None Navigational uses of International Water Courses [115]. This claim is based on the fact that Turkey is reducing the share of other riparian countries by implementing the extensive GAP project. This is damaging the agricultural practices both in Iraq and Syria as well as municipal water and health sectors. Therefore, the project has negative influences on the people and the environment where the quality of water is deteriorating. It is noteworthy to mention that Iraq is the most negatively affected country due to impounding of dams both in Turkey and Syria as well as water quality deterioration.

Finally, it should be stated that Syria and Iraq with the occupation of ISIS parts of these countries, the water issue is not considered as a priority now in view of the present situation.

5 Discussion

Countries in the ME suffer from the water shortage problem. This has caused tension and friction and sometimes escalated leading to war between countries in the region. The Tigris and Euphrates Rivers are considered as a very valuable source of water for the riparian countries within the basins of these rivers. Four main countries (Turkey,

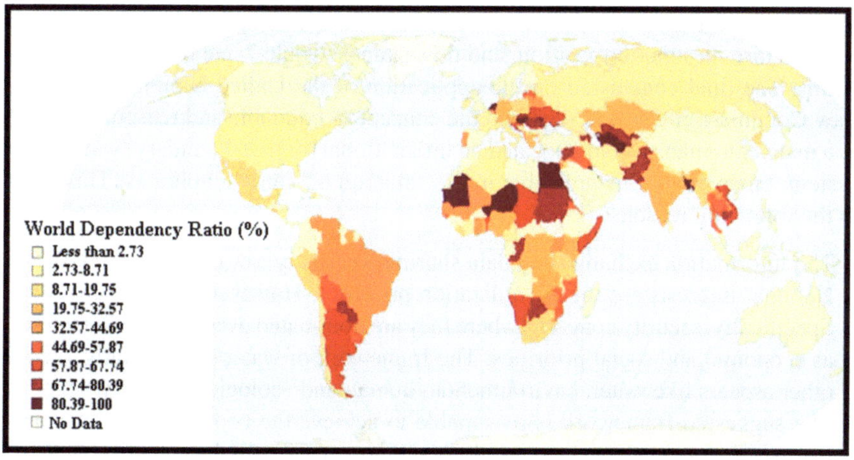

Fig. 11 World dependency ratio (%). Modified after ChartsBin [125]

Syria, Iran and Iraq) utilize the water of Tigris and Euphrates basins and have never reached an agreement to share the waters of these rivers and their tributaries. These countries never reached to an agreement to share the water of these rivers where all of them claim that water is scares and water is over exploited [121–123] or extremely highly stressed [122, 124]. Water Dependency Ratio index shows that Turkey has 1.0% dependence followed by Iran (6.56%), then Iraq (53.45%) and finally Syria (72.36%) (Fig. 11) [125].

Article 6 (1) from the United Nation International Law Commission UN/ILC law reveals that any user of water has exactly the same priority, and this mean that all have equal weight. Using this concept, MacQuarrie [99] analyzed the existing data for the basins of Euphrates and Tigris. The results of the analyses showed that in terms of water needs, Iraq recorded as the lowest between the three countries (1.94). However, slightly lower than Syria by small portion. Therefore, Syria is considered as second (2.0) and Turkey (2.06) is the third. The results revealed that the differences were insignificant that each riparian country had approximately same share from Euphrates waters. According to the fact that equal ranking was used in the analyses, MacQuarrie [99] thinks that this method is not unrealistic because it is an effective analysis tool for some countries. While, it is useless for other countries. Then, MacQuarrie [99] introduced three main factors. The first is priority need, and the second was the Water Security factor, and the third is the environmental factor. As an example of the first factor, MacQuarrie [99] argued that Syria would exceed their social demand, as Turkey's ambitions for use the water resources for generating the energy would overcome its desire for provide water for irrigation. The results of the water security factor which based on take this factor into account depending on the analysis from the previous work showed that all riparian countries have equal ranking. As far as the results of the environmental factor (which was added to the hydrological category) are concerned, he found that as a result of polluted the water of

Euphrates and Tigris from the effluent which is come from agricultural lands, Turkey starts to lose its superior position, and this against Article 7 and the principle of 'no harm'. The final conclusion on the application of the United Nation International Law Commission UN/ILC regarded the concept of equitable and reasonable utilize as a manner to manage and allocated an international (or transboundary) watercourse system, this principal inapplicable in the situation of Tigris–Euphrates. This is due to the following reasons:

- The information exchange and data sharing were very low or does not exist.
- National interests are set the utilization priorities. However, the political issues. Specifically, security concerns where they are dominated over the others issue such as economic and social priorities. The framework of water policies fails to cover other aspects like water, environmental, human and ecological security.
- The suggested framework is reasonable to achieve the cooperation between the involved countries. However, they have no motivation to make it applicable. Especially, Turkey supposed to get losses as Turkey is represent the upstream land. Turkey will not be punished for lack of commitment UN/ILC law because there is no penalty for not meeting it.

In the absence of inclusive agreements and low cooperation, regarding attains the fair sharing for the water resources, individual countries took unilateral steps and implemented projects that led to deterioration the water quality of the whole basin and minimize its local and agricultural benefits. The outcomes of previous negotiations were revealed a few frameworks for basin sharing. However, most of these frameworks stilled without documented agreements. It should be mentioned however, that Turkey exerted some efforts to cooperate. These efforts were primarily concentrated on developing and improving southeastern Turkey (GAP project) where it reduce the dependence on importing oil by seeking for economic development through invest its water abundance [101].

It is noteworthy to mention that there are also political factors and military events that are involved in this conflict [101]. Example of these political factors is the problem between Turkey and Syria because of Hatay Province. France granted Turkey the permission to take control on this province in 1939. In addition, Syria's supports to the PKK [101]. As an example of military events is the conflict that happened in 1975 over the Tabqa Dam, when both Iraqi and Syrian regimes sent military forces to their shared border. In the same contest, Syria granted the PKK part from the Syrian lands, so that they act as a proxy so that they could attack the Turkish locations. Specifically, the strategical hydrological projects. Without any militarily responsibility on Syria [101]. Of course, one should also bear in mind that the eight years' war between Iraq and Iran was because of the sharing of the Shatt Al-Arab watercourse.

It is evident that Turkey's position is strong in comparison with Iraq and Syria as mentioned above that Turkey represent the upstream lands for Tigris and Euphrates Rivers. Due to this fact, as the regional hegemon and the upper riparian, the cooperation is the last thing that Turkey wants. The only factor that makes Turkey accept to negotiate with Syria and Iraq is to avoid any negative criticism from UN or EU and US. Otherwise, Turkish government not serious to cooperate with them. Furthermore, there is no harm by further meetings and discussions. Under the current weak status of Syria and Iraq, it is evident that Turkey is taking advantage and will keep controlling the waters of the Euphrates and Tigris Rivers, and Iran will do the same for the tributaries of the Tigris and Shatt Al-Arab Rivers; Turkey and Iran try to force their plans on Iraq and Syria regardless the consequences. In addition, both Turkey and Iran have the advantage of their geographic position being the upper riparian countries and having relatively the strongest economic and political power in the region which will authorize them to secure any water quantities they desire.

Due to the fact that Syria and Iraq are very much involved in security issues now, Turkey and Iran will remain the riparian hegemons for a long time due to their power and dominant river positions. Water shortage problem is not of prime importance for Syria and Iraq till they solve the ISIS problem but sooner or later, they will try to get their water requirements. To resolve this conflict, a strong and influential mediator is required to bring all parties to the discussion table. Such mediator can be USA, EU or World Bank that can bring the riparian countries to the negotiation table. To give incentives to upper riparian countries to cooperate, other matters can be included in the negotiations such as supplying Turkey with gas and oil from Syria and Iraq for reduced prices. Another important step is required to be implemented by Syria and Iraq is setting a long term strategic plan for the management of their water resources. Such plan is to be implemented irrespective of the changes in the external or internal politics and should be based on "Resources Dependence Theory". This theory assumes that the good human resources, finance and information as well as good international relations exist [126–129].

6 How to Resolve the Conflict?

There are differences in the economy, political and military situations of the riparian countries within the Tigris and Euphrates basins. Recently, Syria has been relatively the weakest countries within the basins due to the ongoing war and security situation. The next weakest country is Iraq. This is due to the outcome of the two Gulf wars and the ongoing war on terrorism. Both countries also are the lowest riparian countries within the Tigris and Euphrates basins. For this reason, Iraq will be the most affected country relative to others. In case the situation remains as it is, all future predictions suggest that all riparian countries will be under water shortage stress [51]. Therefore, quick measures are to be taken to overcome the tension and to resolve any conflict between the riparian countries.

Bilateral talks and/or agreements dominated past efforts to resolve the conflict on water rights, which are not sufficient to begin discussions for a regional solution. Therefore, such negotiations and discussions require a third party to intervene to bring all riparian countries together because they failed to initiate successful triplicate negotiations. Since water issues usually are being eclipsed by more politically charged concerns; then, it is very important to use additional incentives to bring all the parties together [74].

Since Turkey is the riparian hegemon, it should be enticed to negotiate. Two main issues seem to attract the Turkish Government. Finishing the GAP project is the first. This can be done through possible funding agencies like the World Bank or European Union. To do so, Turkey will be asked to sign agreements with downstream riparian counties before releasing any fund. The enticement factors for both Iraq and Syria are development assistance, by financial and technical support and increased water efficiency, which will help in developing their agricultural plans, including more innovative advanced projects. As far as joining the European Union which requires improving the human right issue which can be done by involving the Kurds in a cooperative water utilization effort [74]. To reach a final solution and sign an agreement between riparian countries this requires an external mediator that can highlight and frame the issues in such a way that each country believes that it is gaining by joining the discussion and will lose something by avoiding the discussions.

To achieve this goal, possible mediators can be:

- World Bank: The World Bank has very good technical expertise that can ameliorate the inefficient, water wasting practices of the countries involved. The World Bank can also attract riparian countries by using its financial incentives to reach a resolution so that it can provide loans to these countries. Finally, The World Bank has very good reputation as a mediator in water issues disputes.
- USA and EU: USA and EU possess political, technical, economic powers as well as being international powerful forces and this can be very useful as mediators. All riparian countries would like to get technical and financial support as well as expertise from USA and EU. In addition, Turkey is connected by different treaties with USA (e.g. NATO). As far as EU is concerned, Turkey is trying to be a member within the EU and is also connected with defense agreements.
- United nation: The United Nation has all the required information about the needs and requirements of all riparian countries through its agencies like UNDP, UNEP and FAO. Such information will help in discussions. UN acted as a mediator in different occasions all over the World. Finally, it can use the Security Council if required to enforce agreements.
- Saudi Arabia: This country is influential within the ME region due to its legitimate situation. In addition, it has good financial resources to contribute to a Basin Fund that would finance on going, and future water related plans. Saudi Arabia also has the experience in mediation where it resolved the conflict between Iraq and Syria in 1974–75 where both countries were on the verge of war. It succeeded in bringing the two parties to negotiation and achieved a final resolution to the problem. In 1998, Saudi Arabia acted as a mediator between Turkey and Syria when Turkey accused Syria supporting the PKK and harboring its leader.
- Egypt: Due to the problems concerning the Nile Basin, Egypt gained good experience in discussion and negotiations, which can be used for the Tigris–Euphrates basins.

In addition to the above, riparian countries should set prudent scientific strategic plan to conserve their water resources due to the fact that all these countries will face water shortage problems sooner or later. Al-Ansari [13] set the outlines of such a plan as follow:

6.1 Strategic Water Management Vision

- All official concerned sectors must contribute in planning and designing the "Integrated National Water Master Plan" over long term. The outcomes of this plan should be revealed by the work of the Ministry of Water Resources, Ministry of Municipality and Public Work, Ministry of Agriculture, Water Resources, staff at Universities, private sector, NGO's and representatives of regional and International organizations concerned.
- The hydro infrastructure such as pumping stations, water treatment and hydro power plants should be subjected to a full rehabilitation.
- Encourage the most social sectors to enroll in awareness programs because it is a vital factor to make people to appreciate the seriousness of water issue. In addition, training program for farmers about utilization of new suitable irrigation techniques.
- Defining institutional agenda, including employment and training.
- Supply and demand should be taken into consideration. In this context, seeking for alternative water resources (water harvesting and treated wastewater) beside the conventional methods.
- Promote the private sector to be qualified as an investor in the water sector.
- The coordination between the ministries is quite significant. This will reduce the efforts and save time and money. More decentralization, including budget in irrigation, water supply and sanitation sectors are to be practiced.

6.2 Regional Cooperation and Coordination

- Determine the technical and institutional requirements to build a solid cooperation.
- Collaboration for smooth exchange for transboundary resources. Exert possible efforts for the riparian countries (Turkey, Syria, Iraq and Iran) to attain a reasonable agreement on water quotas.
- United Nation organizations such as UNEP, UNDP, UNESCO, etc., and the International organizations like FAO, WMO, etc. and other research institutions should be requested to share their capabilities in finding solutions for this issue.
- Build communication network with other bodies which they have a good background in water management to get advices in water issues. Specifically, the institutions, organizations and companies in developed countries.

6.3 Irrigation and Agriculture

- Abandoned the conventional irrigation techniques in order to stop to waste the water. On the other hands, adapted new irrigation techniques which they are more compatible for the local conditions of soil, water availability and quality, crops … etc. Apply more efficient methods such as drip irrigation for orchards by using salty water and sprinkler irrigation for grains, and both of them are more conserving than surface irrigation.
- Shifting the conveying system from open channels to closed conduits to increase efficiency of the conveying and minimize the losses. Many advantages could be obtained from closed conduits such as decrease the evaporation and infiltration losses and preventing the direct contact with a saline water table.
- Minimize the soil salinity through improve drainage systems of cultivated lands that will provide better soil leaching. Also, installation a new drainage technique provide an effective solution like use the perforated pipe drainage system in collecting and treating drainage water. Avoiding releases of drainage water to the rivers directly. The path of drainage water should be designed to discharge into the main outfall drain in the areas lying outside the service zone of this drainage project.
- The utilization of chemical fertilizers and pesticides should be reduced. The over use of these materials can cause the deterioration of water quality when the irrigation water release again into the rivers.
- Install modern treatment methods for drainage and sewage water, and reuse it in restoring water bodies, for instance, the marshes.
- Institutions should reflect decentralization, autonomy and farmer empowerment.
- Encourage the investors to take a part in the agricultural sector.
- Set up awareness program to promote the farmers' awareness regards using the suitable techniques in irrigation (drip irrigation and sprinkler irrigation).

6.4 Water Supply and Sanitation

- Improving the efficiency of drinking water distribution networks specially diversion and supply down to the point of use, which is most cost effective.
- Regular maintenance for the pipes of sewerage networks to prevent any leakages that might occur and lead to pollute the surrounding environment.
- Implement new efficient projects to reduce water losses and protect the environment from pollution.
- Improving services, e.g. using Information communication Technology (ICT).
- Expand the sewage network to cover the areas that not serviced with sewage collection services, and connect the newly installed sewage networks to the wastewater treatment plants to avoid the pollution of groundwater from the leakage from old septic tanks.

- Install new sewage treatment plants to satisfy the increased consumption of the domestic sector. Membrane bioreactor technology can be used in these new treatment plants to reuse the treated water.

6.5 *Research and Development*

- Prepare and provide the required information for researchers and decision makers by establishing a comprehensive data base which includes reliable climatological, hydrological, geological, environmental and soil data.
- Motivate the researchers to conduct researches on developing and creating new technologies in water resources and agriculture which more suitable for Iraqi environment.
- Administrate new methods to satisfy our needs from water, and stop our depending on conventional methods. We believe that water harvesting techniques can be very effective and are relative cheap cost wise.
- Raise the level of technicians, engineers and decision makers for understanding the recent utilized technologies through carrying out sessions and training programs.
- Execute pioneer projects, which increase the quantity of water resources, development and productivity, reducing water consumption.
- Placing the structures and outlines of the public awareness programs in terms of water utilization and agricultural activities.
- Asking the universities and institutes to develop their syllabuses to insert special courses in arid region hydrology.
- Awarding of prizes for new innovations, pioneer researches and smart ideas in water resources and their management.
- Exert more efforts to utilize from the groundwater. This source still not exhausted. Therefore, it should be administered wisely and keep it away from the potential contaminant.

References

1. White, G.F.: Resources and Needs: Assessment of the World Water Situation. Water Development, Supply and Management, UK (1978)
2. Gleick, P.H.: Basic water requirements for human activities: meeting basic needs. Water Int. **21**(2), 83–92 (1996). https://doi.org/10.1080/02508069608686494
3. White, G.F.: Introduction world trends and need. Nat. Resour. J. **16**(4), 737–741 (1976)
4. Elhance, A.P.: Hydropolitics in the Third World: Conflict and Cooperation in International River Basins. US Institute of Peace Press, Washington DC, USA (1999)
5. Biswas, A.K.: Water for sustainable development in the twenty-first century: a global perspective. Water Resour. Manage. Ser. **1** (1993)
6. Biswas, A.K.: Management of international waters: problems and perspective. Int. J. Water Resour. Dev. **9**(2), 167–188 (1993)

7. United Nations (UN): International rivers and lakes. In: A Newsletter prepared jointly by the Department for Economic and Social Affairs, United Nations, New York and the Economic Commission for Latin America and the Caribbean, vol. 39. Santiago, Chile (2003)
8. Biswas, A.K.: International Waters of the Middle East from Euphrates-Tigris to Nile (1994)
9. Roger, P., Lydon, P.: Water in the Arab World. Harvard University Press, University Press, Massachusetts, USA (1995)
10. Allan, T.: The ME Water Question. I.B. Tauris Publishers, London (2001)
11. Al-Ansari, N.: Water resources in the Arab countries: problems and possible solutions. In: UNESCO International Conference on World Water Resources at the Beginning of the 21st Century 1998, pp. 3–6
12. Al-Ansari, N.: Management of water resources in Iraq: perspectives and prognoses. Engineering **5**(6), 667–684 (2013)
13. Al-Ansari, N.: Hydro-politics of the Tigris and Euphrates basins. Engineering **8**(3), 140–172 (2016)
14. Cherfane, C.C., Kim, S.E.: Arab region and Western Asia, UNESCWA. In: Managing Water Under Uncertainty and Risk. UN World Water Development Report 4, chap. 33 (2012)
15. Barr, J., Grego, S., Hassan, E., Niasse, M., Rast, W., Talafré, J.: Regional challenges, global impacts. In: Managing Water Under Uncertainty and Risk. UN World Water Development Report 4, Chapter 7 (2012)
16. UN-ESCWA: Inventory of Shared Water Resources in Western Asia. United Nations Economic and Social Commission for Western Asia (2013)
17. Al-Ansari, N., Asaad, N., Walling, D., Hussan, S.: The suspended sediment discharge of the river Euphrates at Haditha, Iraq: an assessment of the potential for establishing sediment rating curves. Geogr. Ann. Ser. A Phys. Geogr. **70**(3), 203–213 (1988)
18. Al-Ansari, N., Toma, A.: Suspended sediment waves in the river Tigris at Sari station for the year 1982. In: International Symposium on Erosion, Debris Flow and Disaster Prevention Keynote, Japan, pp. 157–162 (1985)
19. Al-Ansari, N., Ali, J.: Suspended load and solute discharge in river Tigris within Baghdad. J. Water Res. **5**, 35–50 (1986)
20. Ali, A.: Water Crisis in Iraq: Challenges and Solutions. Al-Bayan Center for Planning and Studies, Baghdad (2018)
21. Ravindranath, A., Devineni, N., Kolesar, P.: An environmental perspective on the water management policies of the upper Delaware river basin. Water Policy **18**(6), 1399–1419 (2016)
22. Najibi, N., Devineni, N., Lu, M.: Hydroclimate drivers and atmospheric teleconnections of long duration floods: an application to large reservoirs in the Missouri river basin. Adv. Water Resour. **100**, 153–167 (2017)
23. Addullah, A.: Shared rivers between Iraq and Iran and its effect on agricultural lands and food security. Tikrit Univ. J. **20**, 356–388 (2012)
24. Wikipedia: List of Dams and Reservoirs in Turkey/Southeastern Anatolia. https://en.wikipedia.org/wiki/List_of_dams_and_reservoirs_in_Turkey (2017)
25. Wikipedia: List of Dams and Reservoirs in Syria. https://en.wikipedia.org/wiki/Category:Dams_in_Syria (2017)
26. Wikipedia: List of dams and reservoirs in Iran. https://en.wikipedia.org/wiki/List_of_dams_and_reservoirs_in_Iran (2017)
27. NASA: Drought in Iraq. https://earthobservatory.nasa.gov/NaturalHazards/view.php?id=38914 (2009)
28. Gleick, P.H.: Water, drought, climate change, and conflict in Syria. Weather Clim. Soc. **6**(3), 331–340 (2014). https://doi.org/10.1175/WCAS-D-13-00059.1
29. Stokes, E.: The drought that preceded Syria's civil war was likely the worst in 900 years. Vice News **3** (2016)
30. Abdullah, A.D.: Modelling Approaches to Understand Salinity Variations in a Highly Dynamic Tidal River: The Case of the Shatt Al-Arab River. CRC Press (2017)
31. Issa, I.E., Al-Ansari, N., Sherwany, G., Knutsson, S.: Expected future of water resources within Tigris-Euphrates rivers basin, Iraq. J. Water Resour. Prot. **6**(5), 421–432 (2014)

32. Al-Shahrabaly, Q.M.: River Discharges for Tigris and Euphrates Gauging Stations. Ministry of Water Resources, Baghdad (2008)
33. Abbas, N., Wasimi, S., Al-Ansari, N., Nasrin Baby, S.: Recent trends and long-range forecasts of water resources of northeast Iraq and climate change adaptation measures. Water **10**(11), 1562 (2018)
34. Al-Ansari, N., Abdellatif, M., Ali, S., Knutsson, S.: Long term effect of climate change on rainfall in northwest Iraq. Open Eng. **4**(3), 250–263 (2014)
35. Al-Ansari, N., Abdellatif, M., Ezz-Aldeen, M., Ali, S.S., Knutsson, S.: Climate change and future long term trends of rainfall at north-east Part of Iraq. J. Civ. Eng. Archit. **8**(6), 790–805 (2014)
36. Al-Ansari, N., Abdellatif, M., Zakaria, S., Mustafa, Y.T., Knutsson, S.: Future prospects for macro rainwater harvesting (rwh) technique in north east Iraq. J. Water Resour. Prot. **6**(5), 403–420 (2014)
37. Al-Ansari, N., Knutsson, S., Almuqdadi, K.: Engineering solution for radioactive waste in IRAQ. J. Adv. Sci. Eng. Res. **4**(1), 18–36 (2014)
38. Najibi, N., Devineni, N.: Recent trends in the frequency and duration of global floods. Earth Syst. Dyn. **9**(2), 757–783 (2018)
39. Altinbilek, D.: Development and management of the Euphrates-Tigris basin. Int. J. Water Resour. Dev. **20**(1), 15–33 (2004)
40. Bilen, Ö.: Turkey and Water Issues in the Middle East. Southeastern Anatolia Project (GAP) Regional Development Administration (1997)
41. Turkish Ministry of Foreign Affairs: Water Issues between Turkey, Syria and Iraq, Department of Transboundary Waters. http://sam.gov.tr/wp-content/uploads/2012/01/WATERISSUES-BETWEEN-TURKEY-SYRIA-AND-IRAQ.pdf, Last accessed January 12, 2018 (2012)
42. Osman, Y., Abdellatif, M., Al-Ansari, N., Knutsson, S., Jawad, S.: Climate change and future precipitation in arid environment of middle east: case study of Iraq. J. Environ. Hydrol. **25**, 1–18 (2017)
43. Osman, Y., Al-Ansari, N., Abdellatif, M.: Climate change model as a decision support tool for water resources management: a case study of greater Zab river. J. Water Clim. Change **8**, 1–14 (2017)
44. IPCC: Contribution of working groups I, II and III to the fourth assessment report of the intergovernmental panel on climate change. In: Climate Change 2007: Synthesis Report, Geneva, Switzerland, p. 104 (2007)
45. Voss, K.A., Famiglietti, J.S., Lo, M., De Linage, C., Rodell, M., Swenson, S.C.: Groundwater depletion in the Middle East from GRACE with implications for transboundary water management in the Tigris-Euphrates-Western Iran region. Water Resour. Res. **49**(2), 904–914 (2013)
46. Chenoweth, J., Hadjinicolaou, P., Bruggeman, A., Lelieveld, J., Levin, Z., Lange, M.A., Xoplaki, E., Hadjikakou, M.: Impact of climate change on the water resources of the eastern Mediterranean and Middle East region: modeled 21st century changes and implications. Water Resour. Res. **47**(6) (2011). https://doi.org/10.1029/2010wr010269
47. Bazzaz, F.: Global climatic changes and its consequences for water availability in the Arab world. In: Roger, R., Lydon, P. (eds.) Water in the Arab Word: Perspectives and Prognoses, pp. 243–252. Harvard University (1993)
48. UNDP (United Nations Development Program): Drought impact assessment. In: Recovery and Mitigation Framework and Regional Project Design in Kurdistan Region (KR). Available at: http://www.undp.org/content/dam/rbas/report/Drought.pdf. (2011)
49. Hameed, M., Ahmadalipour, A., Moradkhani, H.: Apprehensive drought characteristics over Iraq: results of a multidecadal spatiotemporal assessment. Geosciences **8**(2), 58 (2018)
50. United Nations (UN): Water Resources Management White Paper. In: U.N.C.T.i.I. (ed.). United Nations Assistance Mission for Iraq (2010)
51. Maddocks, A., Young, R.S., Reig, P.: Ranking the World's Most Water-Stressed Countries in 2040. http://www.wri.org/blog/2015/08/ranking-world%E2%80%99s-most-water-stressed-countries-2040 (2015)

52. Drake, C.: Water resource conflicts in the Middle East. J. Geogr. **96**(1), 4–12 (1997). https://doi.org/10.1080/00221349708978749
53. Worldmeters: Iran Population. http://www.worldometers.info/world-population/iran-population/ (2018)
54. Worldmeters: Iraq Population. http://www.worldometers.info/world-population/iraq-population/ (2018)
55. Worldmeters: Syria Population. http://www.worldometers.info/world-population/syria-population/ (2018)
56. Worldmeters: Turkey Population. http://www.worldometers.info/world-population/turkey-population/ (2018)
57. Abumoghli, I.: Water Scarcity in the Arab World. http://www.ecomiddleeastandnorthafrica.org/water-arab/ (2015)
58. Hiniker, M.: Sustainable solutions to water conflicts in the Jordan valley. Camb. Rev. Int. Aff. **12**(2), 255–273 (1999). https://doi.org/10.1080/09557579908400261
59. Sadik, A.-K., Barghouti, S.: The water problems of the Arab world: management of scarce water resources. In: Water in the Arab World, pp. 4–37. Harvard University Press, Cambridge, Massachusetts (1994)
60. Charrier, B., Curtin, F.: A vital paradigm shift to maintain habitability in the Middle East: the integrated management of international watercourses. In: Water for Peace in the Middle East and Southern Africa, p. 11 (2007)
61. Hillel, D.: Rivers of Eden: The Struggle for Water and the Quest for Peace in the ME. Oxford University Press and Madzini and Wolf, N.Y. (1994)
62. New York Times: Without Water, Revolution (2013)
63. United Nations and World Bank: Joint Iraq Needs Assessment Working Paper (2003)
64. Al Ansary, K.: Wheat-Importing Iraq Plans to be Net Grain Exporter by 2017 (2015)
65. Robertson, C.: Iraq suffers as the Euphrates river dwindles. N. Y. Times **13** (2009)
66. Cockburn, P.: As Iraq runs dry, a plague of snakes is unleashed. Independent (2009)
67. FAO: AQUASTAT Global Water Information System, Turkey (2008)
68. FAO: AQUASTAT Global Water Information System, Syria (2008)
69. FAO: AQUASTAT Global Water Information System, Iran (2008)
70. FAO: AQUASTAT Global Water Information System, Iraq (2008)
71. FAO: Turkey, Water Report 34 (2009)
72. FAO: Syria, Water Report 34 (2009)
73. FAO: Iraq, Water Report 34 (2009)
74. Akanda, A., Freeman, S., Placht, M.: The Tigris-Euphrates River Basin: Mediating a Path towards Regional Water Stability. Fletcher Sch. J. Issues Relat. Southwest Asia Islam. Civilization (2007)
75. Turan, I.: Water and Turkish foreign policy. In: The Future of Turkish Foreign Policy, pp. 191–208 (2004)
76. Bagis, A.I., Gap, S.A.P.: The Cradle of Civilisation Regenerated. Interbank (1989)
77. Waterbury, J.: Transboundary water and the challenge of international cooperation in the Middle East. In: Water in the Arab World: Perspectives and Prognoses, pp. 39–64 (1994)
78. Alsowdani, M.: GAP project and its economic negative implications on Syria and Iraq. Aletehad News (2005)
79. Shams, S.: Water conflict between Iraq and Turkey. Middle East News. http://www.mokarabat.com/m1091.htm (2006)
80. Alnajaf News net: The GAP project and its negative implications on Iraq. Alnajaf News net (2009)
81. National Defense Magazine: Turkish Israeli partnership in GAP southeastern Anatolian project. Official site of the Lebanese Army, http://www.lebarmy.gov.lb/ article.asp?ln=ar&id=2901 (2009)
82. CEB (Consulting Engineering Bureau): Tigris and Euphrates Sampling. Final Report, College of Engineering, University of Baghdad, Iraq (2011a)

83. World Bank: Iraq: Country Water Resources, Assistance Strategy: Addressing Major Threats to People's Livelihoods, vol. 36297-IQ 97p (2006)
84. IMMPW: Water Demand and Supply in Iraq: Vision, Approach and Efforts, GD for Water. http://www.mmpw.gov.iq/ (2011)
85. Shahin, M.: Water Resources and Hydrometeorology of the Arab Region, vol. 59. Springer Science & Business Media (2007)
86. Taylor and Francis Group: The Europa World Year Book 2003, vol. 1. Taylor & Francis (2003)
87. CEB (Consulting Engineering Bureau): Lakes Testing Study. College of Engineering, University of Baghdad, Baghdad (2011b)
88. Reed, C.: Paradise lost: what should-or can-be done about the environmental crime of the century. Harvard Mag. (2005)
89. Kolars, J.: Problems of international river management : the case of the Euphrates. In: International Waters of the Middle East: From Euphrates-Tigris to Nile, pp. 44–94 (1994)
90. Varela-Ortega, C., Sagardoy, J.A.: Analysis of irrigation water policies in Syria: current developments and future options. In: International Conference on Irrigation Water Policies: Micro and Macro Considerations, Agadir, Morocco, pp. 15–17 (2002)
91. Friedman, T.: Without Water, Revolution (2013)
92. Al-Ansari, N., Knutsson, S.: Possibilities of restoring the Iraqi marshes known as the Garden of Eden. In: Water and Climate Change in the MENA-Region: Adaptation, Mitigation and Best Practices, 28–29 Apr 2011
93. Al-Ansari, N., Knutsson, S., Ali, A.: Restoring the Garden of Eden, Iraq. J. Earth Sci. Geotech. Eng. **2**(1), 53–88 (2012)
94. Hamdy, A.: Water crisis and food security in the Arab world: the future challenges. Global Water Partnership: Mediterranean. http://www.gwpmed.org/files/IWRM-Libya/Atef%20Hamdy%20AWC.pdf (2013)
95. Venter, A.J.: The oldest threat: water in the Middle East. Middle East policy **6**(1), 126–137 (1998)
96. Morris, M.E.: Poisoned Wells: The Politics of Water in the Middle East (2004)
97. Jones, M.: Critical Factors in Transnational River Disputes: an Analytical Framework for Understanding the India-Bangladesh Water Scarcity Dispute Over the Ganges river. The Miami University at Oxford, Ohio (1995)
98. Schaap, W., van Steenbergen, F.: Ideas for Water Awareness Campaigns. GWP (2002)
99. MacQuarrie, P.: Water Security in the ME, Growing Conflict Over Development in the Euphrates-Tigris Basin (2004)
100. Gruen, G.E.: Turkish waters: source of regional conflict or catalyst for peace? In: Environmental Challenges, pp. 565–579. Springer (2000)
101. Berardinucci, J.: The Impact of Power on Water Rights: a Study of the Jordan and Tigris-Euphrates Basins. School of International Service, American University, Washington DC (2010)
102. Beschorner, N.: Water and Instability in the ME. Adelphi Paper 273 (2008)
103. El-Fadel, M., El Sayegh, Y., Ibrahim, A.A., Jamali, D., El-Fadl, K.: The Euphrates-Tigris basin: a case study in surface water conflicts resolution. J. Nat. Resour. Life Sci. Educ. **31**, 99–110 (2002)
104. Kibaroglu, A.: The role of epistemic communities in offering new cooperation frameworks in the Euphrates-Tigris rivers system. J. Int. Aff. **61**(2), 183–198 (2008)
105. Bari, Z.: Syrian-Iraqi dispute over the Euphrates waters. Int. Stud. **16**(2), 227–244 (1977)
106. Kibaroglu, A., Scheumann, W.: Euphrates-Tigris rivers system: political rapprochement and transboundary water cooperation. In: Turkey's Water Policy, pp. 277–299. Springer (2011)
107. Daoudy, M.: Asymmetric power: negotiating water in the Euphrates and Tigris. Int. Negot. **14**(2), 361–391 (2009)
108. Lowi, M.R.: Water and Power: the Politics of a Scarce Resource in the Jordan River Basin, vol. 31. Cambridge University Press (1995)
109. Zawahri, N.A.: Stabilising Iraq's water supply: what the Euphrates and Tigris rivers can learn from the Indus. Third World Q. **27**(6), 1041–1058 (2006)

110. Schmandt, J., Kibaroğlu, A.l.: Sustainability of Engineered Rivers in Arid Lands, PRP 190. LBJ School of Public Affairs (2016)
111. Wolf, A.T., Newton, J.T.: Case Study of Transboundary Dispute Resolution: the Tigris-Euphrates basin, Appendix: C of the book on Transboundary Dispute Resolution by the same authors, Oregon State University; Institute of water and watersheds (2008). Available at: https://www.researchgate.net/publication/237780392_Case_Study_Transboundary_Dispute_Resolution_the_Tigris-Euphrates_basin. Last accessed March 16, 2018
112. Wallace, C.P.: Iran, Iraq still fail to bridge waterway dispute. Los Angeles Times (1998). http://articles.latimes.com/1988-08-19/news/mn-739_1_shatt-al-arab. Last accessed March 19, 2018
113. Guru, D.A.: Iraq-Iran Talks on Borders, Shatt al-Arab, Iraq (2014)
114. ILA, Helsinki: The Helsinki Rules, of the Uses of Waters of International Rivers. Report of the ILA Committee on the use of waters of International Rivers London. Available at: http://www.thehinducentre.com/multimedia/archive/03021/The_Helsinki_Rules_3021443a.pdf. Last visited March 18, 2018 (1967)
115. United Nations General Assembly: Convention on the Law of the Non-Navigational Uses of International Water-courses, Official Records of the General Assembly. In: Session, F.-f. (ed.) vol. 49 (A/51/49). Available at: http://legal.un.org/ilc/texts/instruments/english/conventions/8_3_1997.pdf. Lasted accessed March 18, 2018 (1997)
116. Bremer, N.: Dams on Euphrates and Tigris: impact and regulation through international law, chap. 7. In: Water Law and Cooperation in the Euphrates-Tigris Region, p. 145. Brill (2013)
117. Tomanbay, M.: Turkey's water potential and the southeast Anatolia project. In: Water Balances in the Eastern Mediterranean, pp. 95–112 (2000)
118. Yuksel, I.: South-eastern Anatolia project (GAP) factor and energy management in Turkey. Energy Rep. **1**, 151–155 (2015)
119. Oei, P., Siehlow, M.: Modelling water management optimal for the Tigris-Euphrates river sheds. Int. J. Environ. Sci. **1**, 40–150 (2016). http://www.iaras.org/iaras/filedownloads/ijes/2016/008-0022.pdf
120. Centre for Strategic Research: Water Issues Between Turkey, Syria and Iraq. Turkish ministry of foreign affairs, June-August 1996. Available at: http://sam.gov.tr/water-issues-between-turkey-syria-and-iraq. Last accessed March 18 (1996)
121. Smakhtin, V.: Taking Into Account Environmental Water Requirements in Global-Scale Water Resources Assessments, vol. 2. Iwmi (2004)
122. Aydin, M., Ereker, F.: Water scarcity and political wrangling: security in the Euphrates and Tigris basin. In: Facing Global Environmental Change, pp. 603–613. Springer (2009)
123. UN: Population Dynamics in the Post-2015 Development Agenda, New York (2013)
124. Reig, P., Maddocks, A., Gassert, F.: World's 36 Most Water-Stressed Countries. World Resources Institute. http://www.wri.org/blog/2013/12/world%E280 (2013)
125. ChartsBin: Total Renewable Water Resources Dependency Ratio by Country. http://chartsbin.com/view/1471
126. Pfeffer, J.: A resource dependence perspective on intercorporate relations. In: Intercorporate Relations: the Structural Analysis of Business, pp. 25–55 (1987)
127. Preffer, J., Salancik, G.: The External Control of Organizations: a Resource Dependence Perspective (1978)
128. Hillman, A.J., Withers, M.C., Collins, B.J.: Resource dependence theory: a review. J. Manage. **35**(6), 1404–1427 (2009). https://doi.org/10.1177/0149206309343469
129. Harkins, J., Forster-Holt, N.: Resource dependence and the exits of young firms. Entrep. Res. J. **4**(4), 323 (2014)

A New Ecological Risk Assessment Method of Heavy Metals in Sediments and Soil

Emad Al-Heety

Abstract The aim of this work was to derive a new modified equation to assess the potential ecological risk (E_r) of heavy metals in riverine sediments and soils. The new equation calculates the ecological risk (E_r) in terms of the geoaccumulation index $\left(I_{geo}\right)$. Six new equations were derived to assess the E_r of Cd, Cr, Cu, Ni, Pb and Zn. The E_r of heavy metals in sediments of the Euphrates, Iraq and the Tietê River, Brazil was assessed using the new equation. The E_r was also assessed for the heavy metals in soils of Fallujah, Iraq and Tarkwa, Ghana. Results of application of the new equation were compared with those resulted from common equation (Hakanson's equation). Results of the comparison give credibility to use the new equation for ecological risk assessment. The effect of the reference value and concentration of heavy metal on E_r value was investigated.

Keywords Metal · Pollution · Ecological risk · Soil · Sediment

1 Introduction

Recently, the heavy metal pollution has attracted global concern as serious environmental issue because of its toxicity, bioaccumulation, abundance and persistence [1–3]. The primary sources of heavy metals accumulation in sediments of the aquatic environment are chemical leaching of bedrocks, water drainage basins and runoff from banks [4]. The anthropogenic activities such as mining operations, disposal of industrial wastes and application of pesticides are also heavy metals pollution sources in sediments of the aquatic systems [5]. The polluted sediments, in turn, can act as source of heavy metals conveying them into the water and degrading water quality [6, 7]. Sediments polluted by heavy metals can reflect the water quality of aquatic systems [8]. The main sources of heavy metals accumulation in soils and especially in urban soil are industrial activities, coal and fuel combustion, vehicles emissions,

E. Al-Heety (✉)
Department of Applied Geology, University of Anbar, Rammadi, Iraq
e-mail: emadsalah@uoanbar.edu.iq

Y. T. Mustafa et al. (eds.), *Recent Researches in Earth and Environmental Sciences*, Springer Proceedings in Earth and Environmental Sciences,
https://doi.org/10.1007/978-3-030-18641-8_5

mining operations, fertilizers and pesticides use, municipal solid waste disposal, and other wastes [1, 9]. Heavy metals in sediments and soils can be spread and accumulated in plants and taken by human through consumption. The heavy metals cumulate in greasy tissues and then affect the functions of nerves, endocrine, and immune systems, normal cellular metabolism [10–12]. The ecological risk assessment is tool to some extent specifies the probability of an adverse effect to an organism or ecosystem due to exposure to environmental stressors, such as, chemical or biological pollution [13]. Different methods were developed to calculate the ecological risk of heavy metal, like potential ecological risk index $\left(E_r^I\right)$ [14] and index of geoaccumulation $\left(I_{geo}\right)$ [15]. After the pioneer work of Hakanson's [14], several authors have proposed modified methods that take into consideration the chemical fractions and bioavailability of heavy metals to assess the ecological risk. These modified methods include the risk assessment code (RAC) proposed by Perin et al. [16], the multiparameter evaluation index (MPE) suggested by Thurston and Spengler [17] and the modified ecological risk (MRI) introduced by Kulikowska et al. [18]. I_{geo} and E_r^I are widely employed as a quantitative measure of the potential risk of heavy metals in sediments and soils. These two indices $\left(I_{geo}\text{ and } E_r^I\right)$ have been extensively employed to assess E_r^I of heavy metals pollution in sediments of aquatic systems, [3, 18–24]. Both indices have been also used to evaluate the ecological risk in polluted soils by heavy metals [25–28]. This work aims to: (1) finding relationship between E_r^I and I_{geo}, (2) applying the new relation in different cases study and (3) comparing the results of application of the new modified and Hakanson methods.

2 Methodology of Ecological Risk Assessment

Two methods were widely employed as a quantitative measure of the potential ecological risk level of heavy metals in sediments and soils. These two methods are I_{geo} and E_r^I. I_{geo} is determined using the following equation [15]:

$$I_{geo} = \log_2\left(C_s^i / K C_r^i\right) \tag{1}$$

$$I_{geo} = \log_2\left(C_s^i / 1.5 C_r^i\right) \tag{2}$$

where C_s^i refers to the concentration of heavy metal i in the sample and C_r^i is the reference value of heavy metal i. K is a constant in view of the reference value fluctuation. The factor 1.5 is introduced as a value of K constant to include the possible variation of the reference values due to lithogenic effect. Muller [15] classified the I_{geo} values as follows: (0) practically unpolluted ($I_{geo} \leq 0$), (1) unpolluted to moderately polluted ($0 < I_{geo} \leq 1$), (2) moderately polluted ($1 < I_{geo} \leq 2$), (3) moderately to heavily polluted ($2 < I_{geo} \leq 3$), (4) heavily polluted ($3 < I_{geo} \leq 4$), (5) heavily to extremely polluted ($4 < I_{geo} \leq 5$), (6) extremely polluted ($5 < I_{geo}$).

The E_r^I was suggested by Hakanson [14] and used to assess the ecological risk of heavy metal in sediments and soils. The E_r^I is calculated using the following equations:

$$E_r^i = T_f^i \times C_f^i \tag{3}$$

$$C_f^i = C_s^i / C_r^i \tag{4}$$

$$E_r^i = T_f^i \left(C_s^i / C_r^i \right) \tag{5}$$

where E_r^i is the potential ecological risk factor of metal i, T_f^i is the toxic response factor of metal i. The values of heavy metals are 30, 2, 5, 5, 5, and 1 for Cd, Cr, Cu, Ni, Pb and Zn, respectively. C_s^i is the metal i concentration in a sediment or soil sample, and C_r^i is the reference value of metal i. E_r^i results are classified as follows: low potential ecological risk ($E_r^i < 40$), moderate potential ecological risk ($40 \leq E_r^i < 80$), considerable potential ecological risk ($80 \leq E_r^i < 160$), high potential ecological risk ($160 \leq E_r^i < 320$), and very high ecological risk ($320 \leq E_r^i$).

2.1 New Modified Equation for Assessment of E_r^i

Based on the methods mentioned above, the E_r^i depends on the concentration and the toxicity response factor of heavy metal. In Eqs. (2) and (5) C_s^i / C_r^i are common in both equations. In term of this common limit, new modified equation to assess the ecological risk was derived as follow:

Since $C_f^i = C_s^i / C_r^i$, (2) becomes

$$I_{geo} = \log_2 \left(\frac{2}{3} \times c_{f^i} \right) \tag{6}$$

Rewriting Eq. (3)

$$C_{f^i} = \frac{E_r^i}{T_f^i} \tag{7}$$

Combination of (6) and (7), we obtain

$$I_{geo} = \log_2 \left(\frac{2}{3} \times \frac{E_r^i}{T_f^i} \right) \tag{8}$$

According to the logarithmic rules, we rewrite (8)

$$I_{geo} = \log_2 \frac{2}{3} + \log_2 \frac{E_r^i}{T_f^i} \tag{9}$$

$$I_{geo} = 1 - 1.58 + \log_2 E_r^i - \log_2 T_f^i \tag{10}$$

Then

$$\log_2 E_r^i = I_{geo} + \log_2 T_f^i + 0.58 \tag{11}$$

Substitution value of T_f^i for each metal, we obtain the following equations:

$$\log_2 E_r^{cd} = 5.486 + I_{geo} \tag{12}$$

$$\log_2 E_r^{cr} = 1.58 + I_{geo} \tag{13}$$

$$\log_2 E_r^{cu} = 2.901 + I_{geo} \tag{14}$$

$$\log_2 E_r^{Ni} = 2.901 + I_{geo} \tag{15}$$

$$\log_2 E_r^{Pb} = 2.901 + I_{geo} \tag{16}$$

$$\log_2 E_r^{Zn} = 0.58 + I_{geo} \tag{17}$$

Using of the equations mentioned above, we can calculate Er for *Cd, Cr, Cu, Ni, Pb* and *Zn* in term of I_{geo}. The difference between our new modified method and the classical Hakanson's method is in calculation of pollution index C_f^i. In our method, we used the I_{geo} to estimate the pollution index because it takes the possible variation of reference values due to the lithogenic source into consideration.

2.2 Application of the New Modified Equations

The E_r^I of heavy metals (Cd, Cr, Cu, Ni, Pb and Zn) in riverine sediments and soils of four cases was assessed using the new modified equation. The selected cases for sediment of the rivers are of Euphrates River, Iraq and the Tietê River, Brazil, respectively [29, 30]. For soils, the cases are of the urban soil in Fallujah City, Iraq [31] and of the agricultural soils in Tarkwa, Ghana [32]. The heavy metals concentrations (Cd, Cr, Cu, Ni, Pb, and Zn) for each case were listed in Appendix 1. The United States Environmental Protection Agency (USEPA) guidelines for these metals are employed as reference value [33]. These reference values (Cr) are 0.6, 25, 16, 16,

40, and 90, for Cd, Cr, Cu, Ni, Pb, and Zn, respectively. The first step in application of the new modified equation is calculation of I_{geo} for each metal and the second step is using one of the equations for each metal.

3 Results and Discussion

New modified equations were mathematically derived for assessing the E_r^I of some heavy metals in sediments and soil are listed in Table 1. These equations are estimated E_r^I in terms of I_{geo}.

The results of application of the new modified equation for assessing of E_r^I are listed in Appendix 2. To detect if there is relationship between the new modified equation and the original Hakanson's equation, the E_r^I values were also calculated for two cases (sediment of Euphrates River, Iraq and soils in Tarkwa, Ghana) using the Hakanson's equation. A regression analysis between $E_{r(\text{Modified})}$ index and $E_{r(\text{Hakanson})}$ index values was carried out (see Figs. 1 and 2). The result of comparison shows a very significant correlation between the E_r^I values calculated using the new modified equation and those estimated by the Hakanson's equation. A scatterplot between E_r^I values calculated by Hakanson's method and I_{geo} for two cases (sediment of Euphrates River, Iraq and soil of Tarkwa, Ghana) was conducted and listed in Table 2, Appendix 3. The results show significant empirical relations between E_r^I and I_{geo}.

The E_r^I takes into account the metal concentration, reference metal value and toxic response factor of the metal. The toxic response factor is the primary requirement for assessing the E_r^I. The other factor controlling the ecological risk assessment is the selected reference value Cr of the metal. The E_r^I of heavy metals in soil of Fallujah, Iraq was calculated using different reference values. The obtained results show that using different reference values could cause an overestimation or underestimation of the E_r^I. Protano et al. [34] found that the lack of abundance of updated reference metal values could lead to overestimation or underestimation of E_r^I. The accurate

Table 1 The mathematical derived relations and empirical relations for assessing of potential ecological risk index E_r^I of some heavy metals in sediment and soil

Mathematical relations	Empirical relations	
	Sediment of Euphrates river, Iraq	Soil of Tarkwa city, Gahna
$\log_2 E_r^{Cd} = 5.486 + I_{geo}$	$\log_2 E_r^{Cd} = 5.481 + 1.008I_{geo}$	$\log_2 E_r^{Cd} = 5.477 + 0.992I_{geo}$
$\log_2 E_r^{Cr} = 1.580 + I_{geo}$	$\log_2 E_r^{Cr} = 1.581 + 1.001I_{geo}$	$\log_2 E_r^{Cr} = 1.585 + 0.999I_{geo}$
$\log_2 E_r^{Cu} = 2.901 + I_{geo}$	$\log_2 E_r^{Cu} = 2.917 + 0.999I_{geo}$	$\log_2 E_r^{Cu} = 2.971 + 1.015I_{geo}$
$\log_2 E_r^{Ni} = 2.901 + I_{geo}$	$\log_2 E_r^{Ni} = 2.908 + 0.999I_{geo}$	$\log_2 E_r^{Ni} = 2.900 + 0.996I_{geo}$
$\log_2 E_r^{Pb} = 2.901 + I_{geo}$	$\log_2 E_r^{Pb} = 2.903 + 0.999I_{geo}$	$\log_2 E_r^{Pb} = 2.908 + 0.998I_{geo}$
$\log_2 E_r^{Zn} = 0.580 + I_{geo}$	$\log_2 E_r^{Zn} = 0.584 + 0.998I_{geo}$	$\log_2 E_r^{Zn} = 0.582 + 0.997I_{geo}$

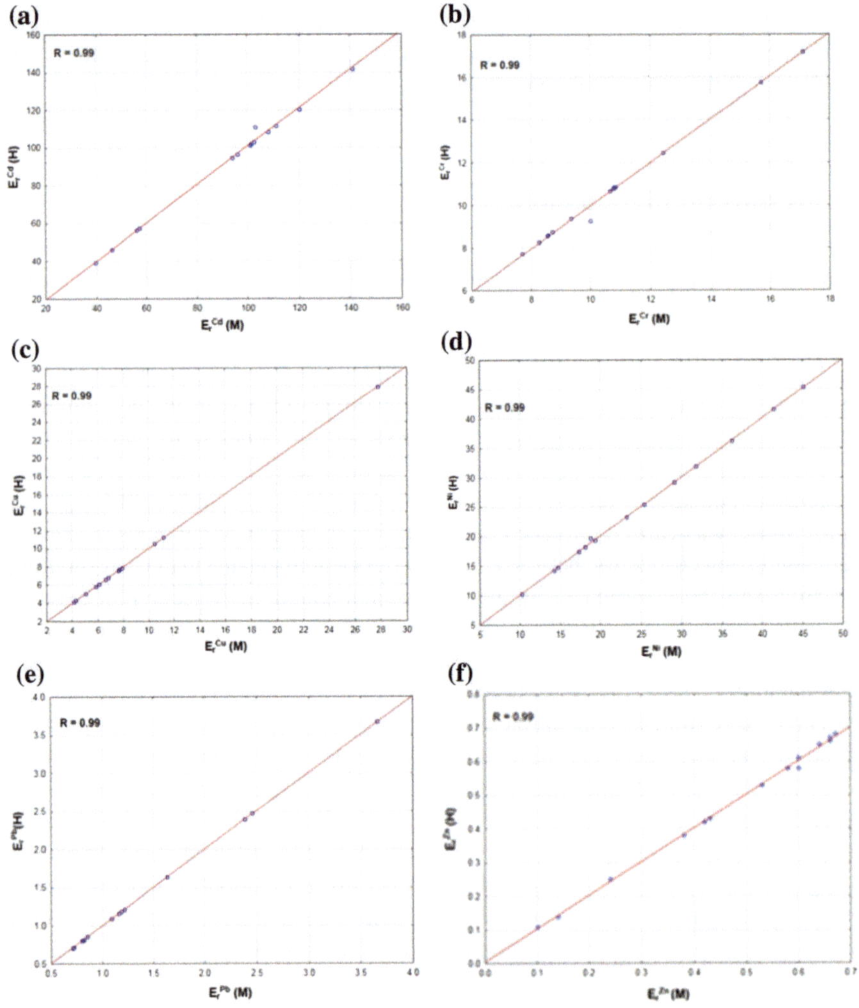

Fig. 1 Scatterplot between the E_r^I values of heavy metals in sediment of Euphrates river, Iraq calculated by Hakanson equation E_r(H) and those calculated by the new modified equation E_r(M)

evaluation of the E_r^I requires regular updating of the reference values periodically at the regional scales, particularly in geologic regions including sensitive ecological habitats [35]. Significant relationships between E_r^I and I_{geo} for Cd, Cr, Cu, Ni, Pb and Zn were recorded for two cases (sediment of Euphrates River, Iraq and soil of Tarkwa City, Gahna), Table 1 and Appendix 3. A significant relation between E_r^I and I_{geo} for Antimony (Sb) was reported in XKS mine, Hunan province in China [36]. The obtained results show significant agreement between the mathematical derived relations and the empirical relations. This agreement and the strong correlation between the E_r^I values calculated by our modified method and those estimated

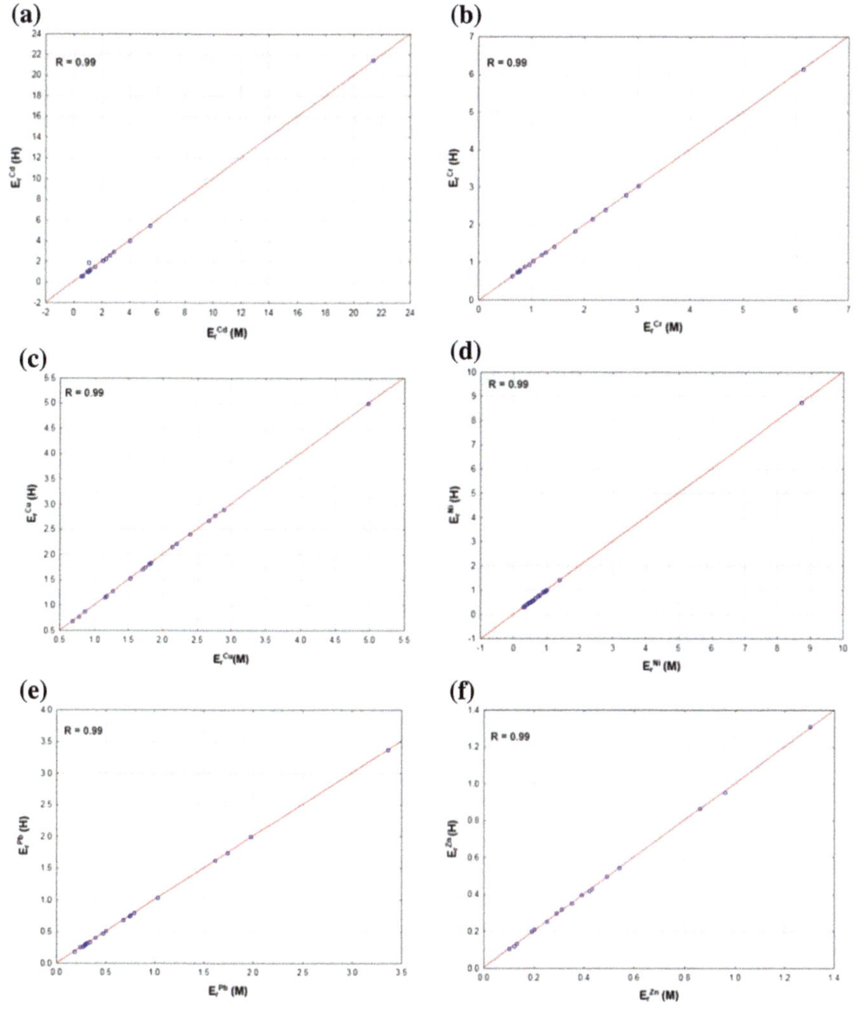

Fig. 2 Scatterplot between E_r^I values of heavy metals in soil of Tarkwa, Ghana calculated by Hakanson equation $E_r(H)$ and those calculated by the new modified equation $E_r(M)$

by Hakanson's method confirms that our method is a reliable approach for assessing the potential ecological risk levels of heavy metals in sediment and soil.

In terms of the total metal concentration, the new modified method suggests more representative values of E_r^I of heavy metals in sediment and soil because it takes into account effect of the possible variation of the reference values due to the lithological sources. In our method, we do not need to know the toxicity response factor values of Cd, Cr, Cu, Ni, Pb and Zn that represent the basic request to calculate E_r^I according to Hakanson's method.

The relationship between total metal concentration, reference metal value and the toxicity response factor of different metals are considered in Hakanson's and our modified methods. However, the chemical fractions concentrations of metals in sediment and soil are not taken into account in both methods. Comparison of the results of assessment of PERI (Hakanson's method) and MRI for some metals in Daya Bay, South China Sea showed that the MEI and MRI values were higher than E_r^I and RI [22]. They interpreted that in terms of modified index of heavy metal concentration. On contrary, using total metal concentration rather than metal fractions contents may lead to an overestimation of the potential ecological risk (PERI) levels [37, 38]. The current study has theoretical and empirical implications and suggests that our new modified method is reliable technique to quantify the ecological risk levels of heavy metals in terms of geo-accumulation index I_{geo}.

4 Conclusions

From the obtained results, the following conclusions were drawn:

- A new modified equation was derived to assess the E_r^I in soil and sediment polluted by heavy metals in terms of geoaccumulation index I_{geo}.
- The results of application of the new modified equation to assess E_r^I were consistent with the results of application of Hakanson's equation.
- The significant empirical relations between E_r^I nd I_{geo} confirmed the mathematical derived relations and inturn gave them credibility to use as an alternative method of Hakanson's method.

5 Appendices

5.1 *Appendix 1*

Heavy metals concentrations (mg/kg) in sediments of Euphrates river, Iraq

Sampling station	Cd	Cr	Cu	Ni	Pb	Zn
S1	0.78	135.8	36	102.21	6.5	38
S2	1.89	135.6	33.7	145.2	9.8	34.5
S3	2.23	103.6	24.3	58.3	6.9	48
S4	2.03	107.5	21.1	74.7	9.35	38.9
S5	2.17	116.1	16.2	32.8	6.63	22.5
S6	2.04	133.6	89.3	47.2	13.15	60.09

(continued)

(continued)

Sampling station	Cd	Cr	Cu	Ni	Pb	Zn
S7	2.22	155.9	13.8	55.6	9.55	12.7
S8	2.06	107.2	21.9	133.2	6.55	53.06
S9	0.92	197	25.3	116	5.83	60.5
S10	2.41	135	19.5	45.6	5.75	40.5
S11	2.83	109.4	18.7	62.8	19.12	58.5
S12	1.13	96.6	13.3	93.6	19.18	52.5
S13	1.15	117.4	18.6	62	8.73	61.2
S14	1.93	214.7	24.7	81.6	16.02	55

Heavy metals concentrations (mg/kg) in sediments of Tietê river, Brazil

Sampling station	Cd	Cr	Cu	Ni	Pb	Zn
S1	1.12	78.3	15.2	18.4	101.8	52.1
S2	3.2	40.8	7.3	16.6	50.9	38.7
S3	5.1	251.8	33.6	29.6	110.6	201.6
S4	5.6	126.8	202.1	72	67.3	421.3
S5	7.5	132.1	73.1	47.9	92.1	218.6
S6	8.5	137.9	115.6	50.3	67.5	390.9
S7	6.3	88.9	39.0	24.7	36.9	108.7
S8	6.3	151.8	76.4	41.0	63.6	107.1
S9	6.8	272.8	144.2	99.6	64.1	164.1
S10	5.4	241.9	56.5	69.4	53.9	57.9
S11	6.4	343.9	359.7	118.2	44.7	150.2
S12	3.5	388.5	293.9	121.9	25.6	181.3

Heavy metals concentrations (mg/kg) in soil of Fallujah city, Iraq

Sampling station	Cd	Cr	Cu	Ni	Pb	Zn
S1	0.800	21.225	2.050	12.050	4.625	8.925
S2	0.575	12.325	0.925	5.575	3.475	3.800
S3	0.825	14.300	1.325	10.475	4.575	5.175
S4	0.650	16.725	1.050	10.375	3.975	5.350
S5	0.575	14.900	1.425	10.525	3.000	6.950
S6	0.600	9.450	1.450	6.400	3.150	3.225

(continued)

(continued)

Sampling station	Cd	Cr	Cu	Ni	Pb	Zn
S7	0.750	11.600	0.875	8.400	2.675	4.750
S8	0.625	11.300	1.000	7.375	2.900	6.425
S9	0.525	10.725	1.450	7.525	2.625	4.925
S10	0.525	9.400	1.000	6.275	2.750	2.750
S11	0.575	9.350	1.600	7.550	3.550	3.100
S12	0.475	7.900	2.175	7.475	3.300	3.800
S13	0.775	10.100	3.825	11.800	4.350	3.825
S14	0.825	9.975	3.050	12.225	4.750	8.450
S15	0.875	9.425	2.325	11.775	4.050	6.775
S16	0.550	9.450	1.425	7.750	3.725	4.500
S17	0.575	12.25	2.500	10.025	4.400	5.125
S18	0.525	9.550	3.125	10.675	4.700	6.775
S19	0.750	11.675	4.975	8.150	5.300	10.400
S20	0.575	10.275	2.750	6.876	4.600	5.000

Heavy metals concentrations (mg/kg) in soil of Tarkwa city, Ghana

Sampling station	Cd	Cr	Cu	Ni	Pb	Zn
S1	0.038	35	8.9	4.5	6.1	39
S2	0.020	30	9.3	2.6	3.2	12
S3	0.011	13	2.8	1.1	1.5	9.7
S4	0.022	9.9	2.8	1.5	2.3	23
S5	0.020	11	5.5	3.1	2.7	32
S6	0.011	15	2.5	1.6	2.0	11
S7	0.020	27	2.2	1.0	2.1	29
S8	0.021	15	3.7	3.2	2.5	19
S9	0.013	9.6	5.9	1.3	2.4	27
S10	0.030	38	5.8	2.4	8.3	36
S11	0.024	9.2	3.8	1.8	3.8	38
S12	0.011	12	5.6	2.1	4.1	18
S13	0.052	8	4.1	1.3	13	86
S14	0.042	12	6.9	1.9	5.5	78
S15	0.081	23	8.6	3.2	16	49
S16	0.11	16	7.1	3	27	78
S17	0.058	18	7.7	2.9	6	32
S18	0.046	12	4.9	1.9	6.4	45
S19	0.43	77	16	28	14	118

5.2 Appendix 2

Ecological risk index values E_r^I of heavy metals in sediments of Euphrates river, Iraq

Sampling station	E_r^I					
	Cd	Cr	Cu	Ni	Pb	Zn
S1	39.72	10.82	11.19	31.8	0.8	0.42
S2	94.09	10.8	10.48	45.16	1.21	0.38
S3	110.96	8.25	7.55	18.13	0.85	0.53
S4	101.05	8.56	6.56	23.23	1.15	0.43
S5	108	9.99	5.04	10.19	0.82	0.24
S6	101.54	10.64	27.78	14.68	1.63	0.66
S7	103.03	12.42	4.29	17.29	1.18	0.14
S8	102.53	8.54	6.78	41.44	0.81	0.6
S9	45.97	15.7	7.86	36.25	0.72	0.66
S10	119.92	10.75	6.06	14.18	0.71	0.450
S11	140.84	8.71	5.81	18.73	2.39	0.64
S12	56.21	7.7	4.13	29.12	2.46	0.58
S13	57.2	9.35	5.79	19.29	1.08	0.67
S14	96.06	17.11	7.68	25.38	3.66	0.6

Ecological risk index values E_r^I of heavy metals in sediments of Tietê river, Brazil, Iraq

Sampling station	E_r^I					
	Cd	Cr	Cu	Ni	Pb	Zn
S1	59.71	6.24	4.73	4.39	12.66	0.57
S2	159.23	3.25	2.27	5.16	6.33	0.42
S3	253.87	20.07	10.45	9.2	13.67	2.23
S4	278.78	10.1	62.85	22.39	8.36	4.66
S5	373.25	10.52	22.73	14.88	11.46	2.41
S6	423.14	10.98	35.95	15.63	8.39	4.32
S7	313.64	7.08	12.13	8.15	4.59	1.2
S8	313.64	12.1	23.76	12.75	7.91	1.18
S9	338.49	21.73	44.84	30.99	7.97	1.81
S10	268.72	19.27	17.58	21.58	6.7	0.63
S11	318.68	27.39	111.89	36.78	5.56	1.67
S12	174.24	30.95	91.45	37.92	3.18	1.35

Ecological risk index values E_r^I of heavy metals in soil of Fallujah city, Iraq

Sampling station	E_r^I					
	Cd	Cr	Cu	Ni	Pb	Zn
S1	39.8	1.69	0.63	3.75	0.57	0.095
S2	28.6	0.98	0.28	1.73	0.42	0.041
S3	41.06	1.14	0.41	3.25	0.56	0.056
S4	32.37	1.33	0.32	3.22	0.49	0.058
S5	28.6	1.19	0.44	3.27	0.37	0.076
S6	29.85	0.75	0.45	1.98	0.38	0.034
S7	37.34	0.92	0.27	2.61	0.32	0.052
S8	31.12	0.90	0.30	2.29	0.35	0.070
S9	26.13	0.85	0.45	2.33	0.32	0.053
S10	26.13	0.74	0.30	1.95	0.33	0.029
S11	28.6	0.74	0.49	2.34	0.44	0.032
S12	23.62	0.62	0.67	2.32	0.41	0.041
S13	38.61	0.80	1.18	3.66	0.53	0.042
S14	41.06	0.79	0.94	3.80	0.59	0.092
S15	43.59	0.75	0.71	3.66	0.50	0.074
S16	27.39	0.75	0.44	2.40	0.46	0.049
S17	28.6	0.97	0.77	3.11	0.54	0.055
S18	26.13	0.75	0.97	3.31	0.58	0.074
S19	37.34	0.92	1.54	2.53	0.65	0.115
S20	28.6	0.82	0.85	2.13	0.56	0.055

Ecological risk index values E_r^I of heavy metals in soil of Tarkwa city, Ghana

Sampling station	E_r^I					
	Cd	Cr	Cu	Ni	Pb	Zn
S1	1.07	2.78	2.76	1.39	0.75	0.43
S2	0.98	2.39	2.89	0.8	0.39	0.13
S3	0.53	1.03	0.86	0.33	0.18	0.1
S4	1.07	0.78	0.86	0.46	0.28	0.25
S5	0.98	0.87	1.71	0.96	0.33	0.35
S6	0.53	1.19	0.77	0.49	0.24	0.12
S7	0.98	2.15	0.68	0.3	0.26	0.31
S8	1.03	1.19	1.15	0.99	0.3	0.2
S9	0.62	0.76	1.83	0.4	0.29	0.29
S10	1.47	3.02	1.8	0.74	1.03	0.39

(continued)

(continued)

Sampling station	E_r^I					
	Cd	Cr	Cu	Ni	Pb	Zn
S11	1.16	0.73	1.18	0.56	0.47	0.42
S12	0.56	0.95	1.74	0.65	0.5	0.19
S13	2.55	0.63	1.27	0.4	1.61	0.96
S14	2.06	0.95	2.14	0.59	0.68	0.86
S15	4.03	1.83	2.67	0.99	1.98	0.54
S16	5.46	1.27	2.2	0.93	3.36	0.86
S17	2.86	1.43	2.39	0.89	0.74	0.35
S18	2.28	0.95	1.52	0.59	0.79	0.49
S19	21.39	6.13	4.97	8.7	1.74	1.3

5.3 Appendix 3

Regression analysis between E_r^i and I_{geo} of Cd, Cr, Cu, Ni, Pb and Zn in sediment of Euphrates River, Iraq.

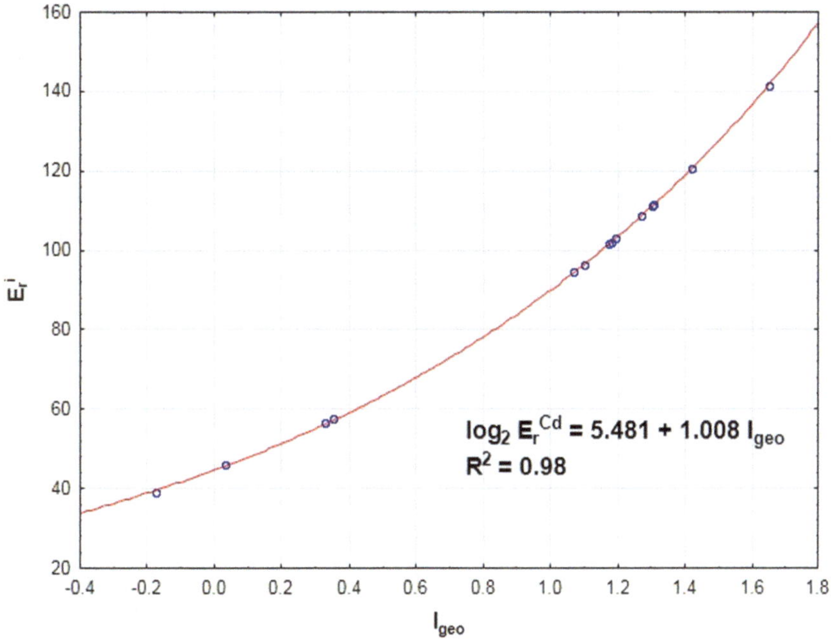

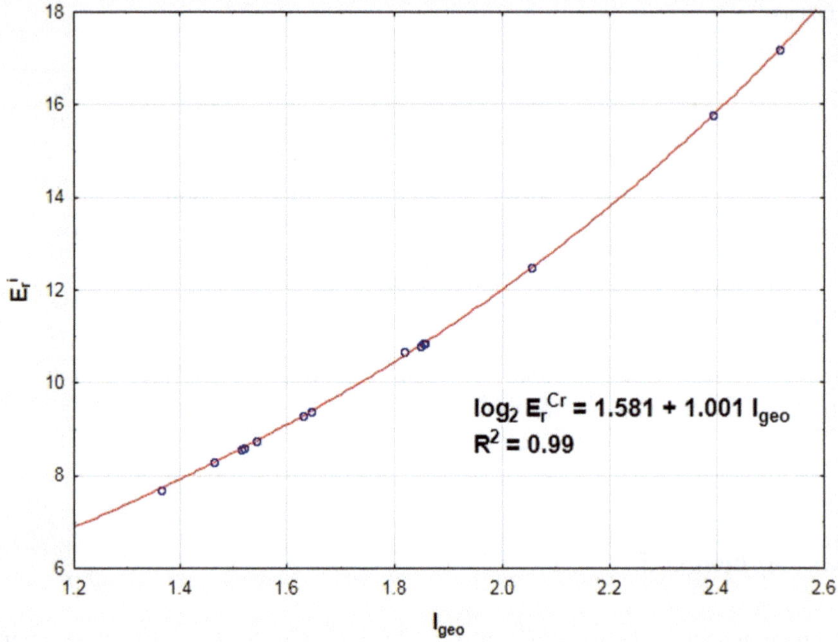
18
16
14
12
10
8
6
1.2
1.4
1.6
1.8
2.0
2.2
2.4
2.6
E_r^i
I_{geo}
$\log_2 E_r^{Cr} = 1.581 + 1.001\ I_{geo}$
$R^2 = 0.99$

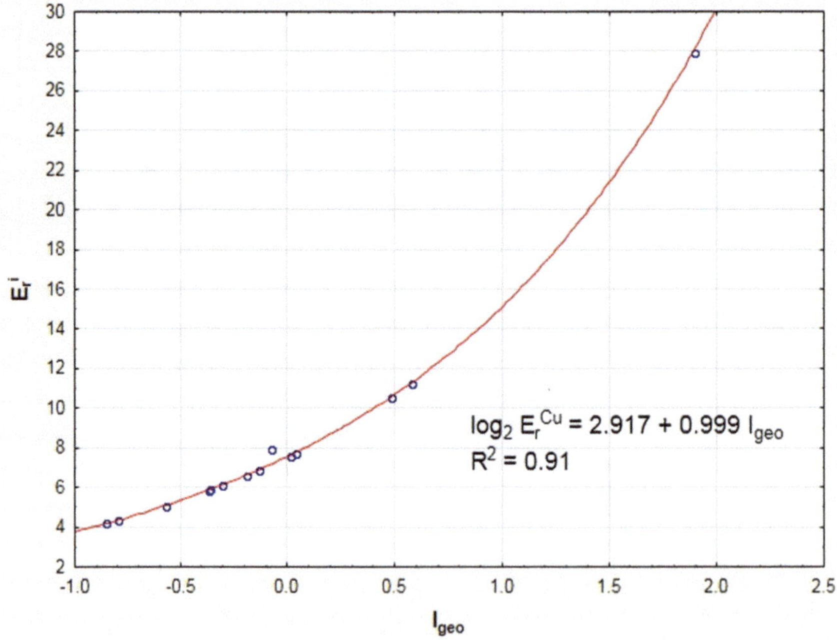
30
28
26
24
22
20
18
16
14
12
10
8
6
4
2
-1.0
-0.5
0.0
0.5
1.0
1.5
2.0
2.5
E_r^i
I_{geo}
$\log_2 E_r^{Cu} = 2.917 + 0.999\ I_{geo}$
$R^2 = 0.91$

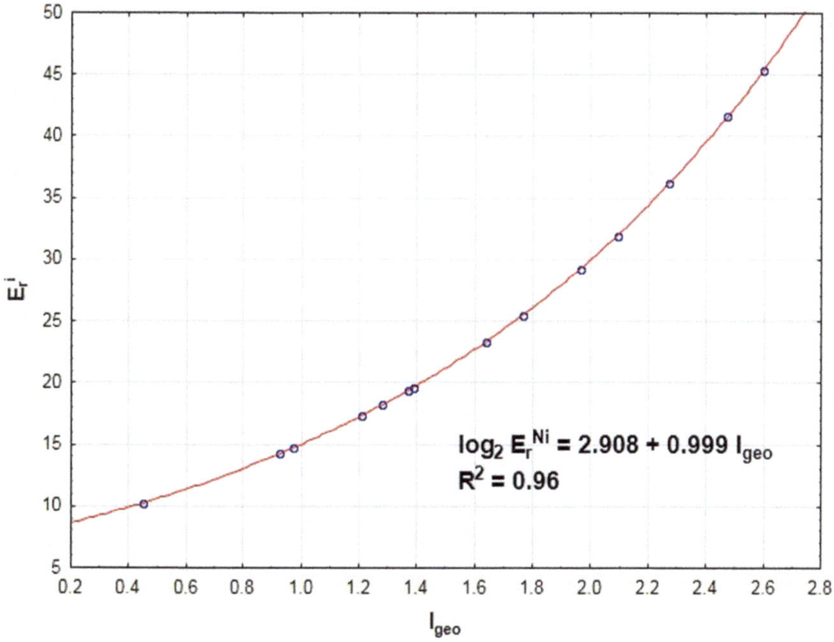
log2 Er^Ni = 2.908 + 0.999 Igeo
R² = 0.96
Er^i
Igeo
50
45
40
35
30
25
20
15
10
5
0.2
0.4
0.6
0.8
1.0
1.2
1.4
1.6
1.8
2.0
2.2
2.4
2.6
2.8

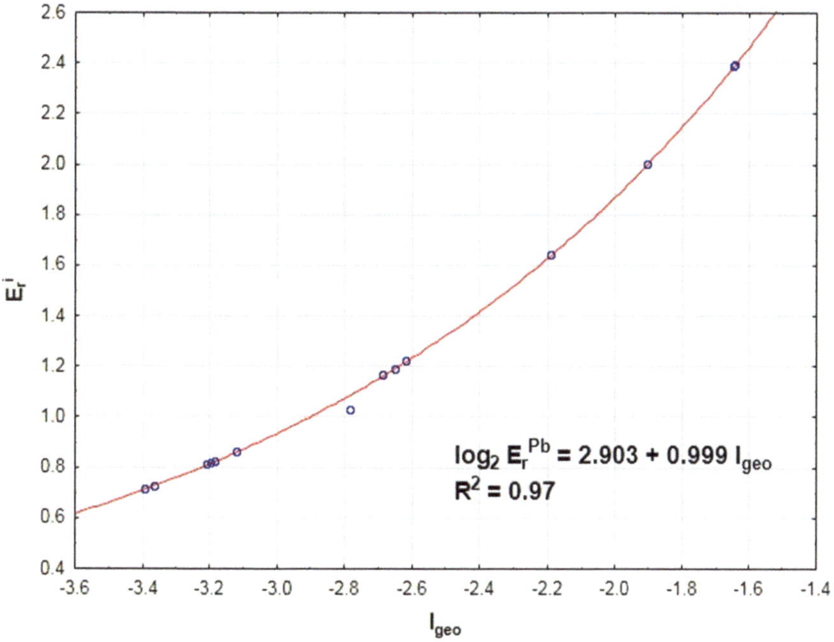
log2 Er^Pb = 2.903 + 0.999 Igeo
R² = 0.97
Er^i
Igeo
2.6
2.4
2.2
2.0
1.8
1.6
1.4
1.2
1.0
0.8
0.6
0.4
-3.6
-3.4
-3.2
-3.0
-2.8
-2.6
-2.4
-2.2
-2.0
-1.8
-1.6
-1.4

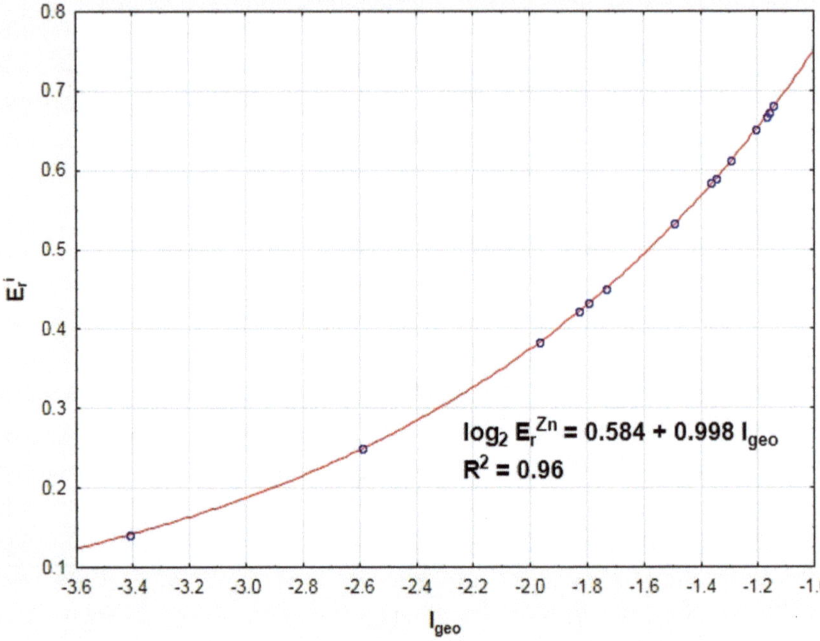

Regression analysis between E_r^I and I_{geo} of Cd, Cr, Cu, Ni, Pb and Zn in soil of Tarkwa, Ghana.

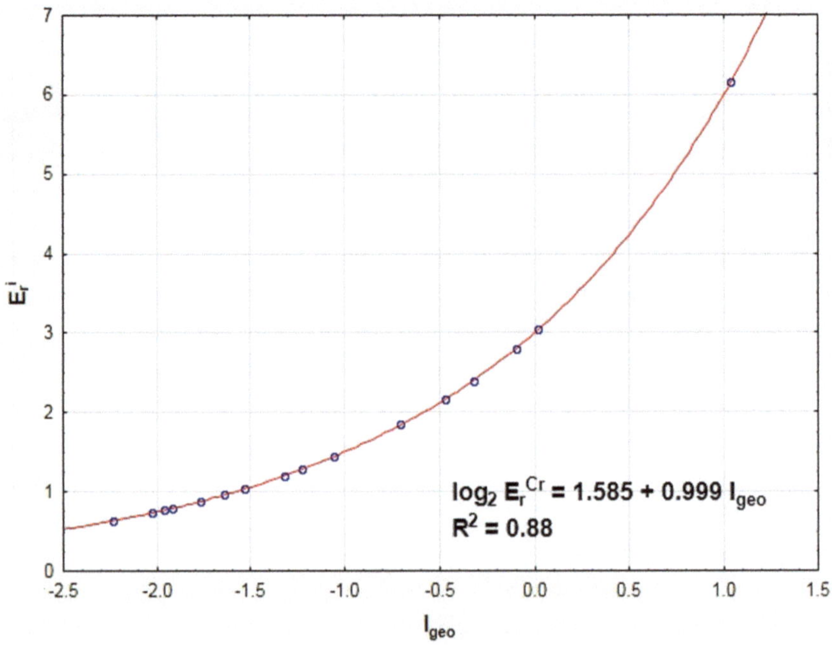

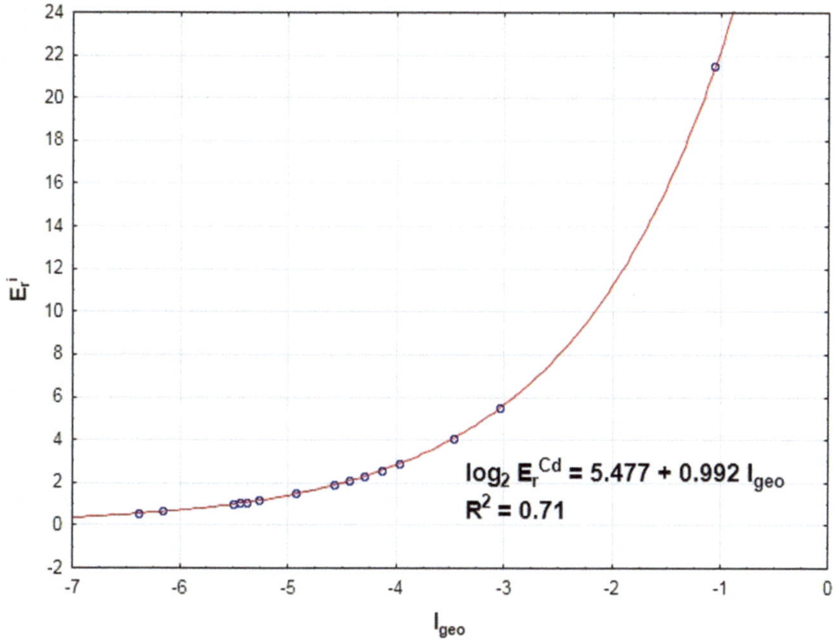
$\log_2 E_r^{Cd} = 5.477 + 0.992\, I_{geo}$
$R^2 = 0.71$
E_r^i
I_{geo}

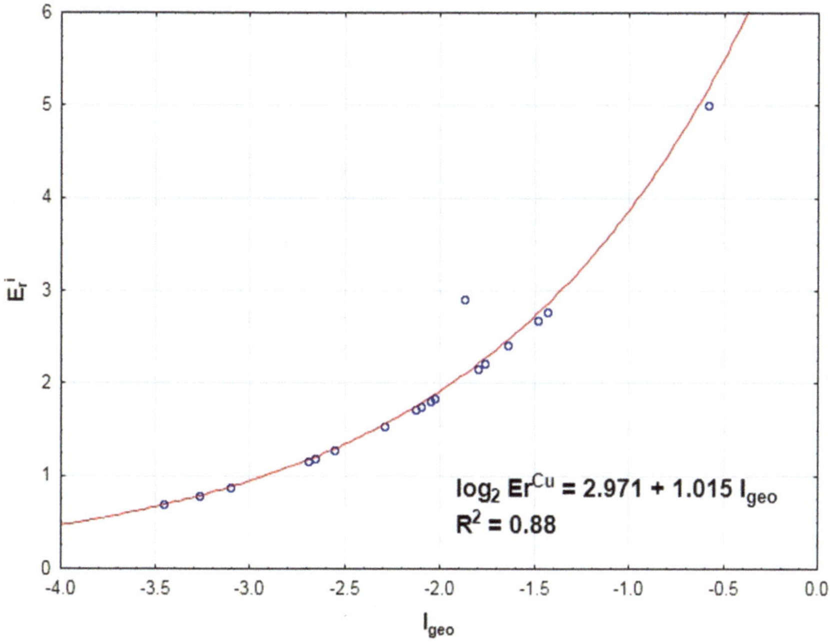
$\log_2 Er^{Cu} = 2.971 + 1.015\, I_{geo}$
$R^2 = 0.88$
E_r^i
I_{geo}

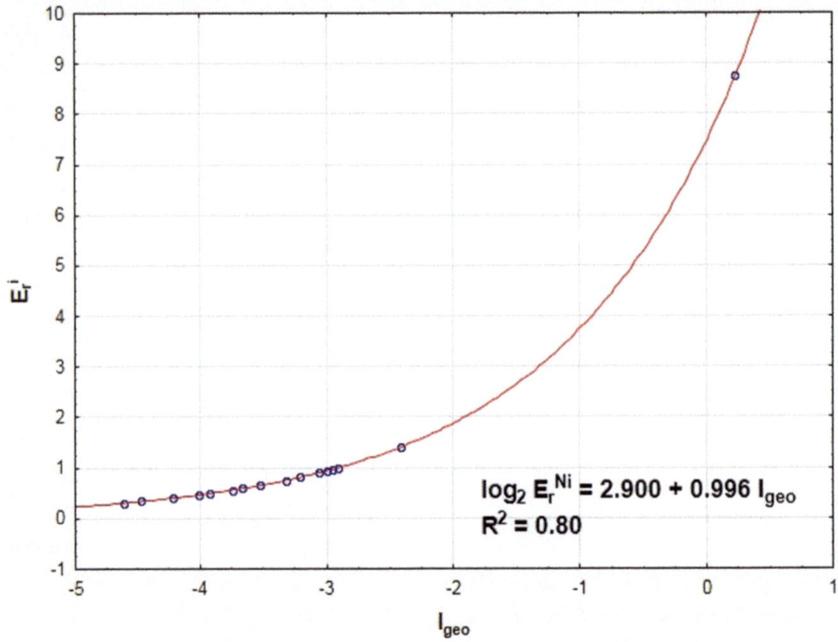

log2 Er^Ni = 2.900 + 0.996 Igeo
R2 = 0.80
Er^i
Igeo

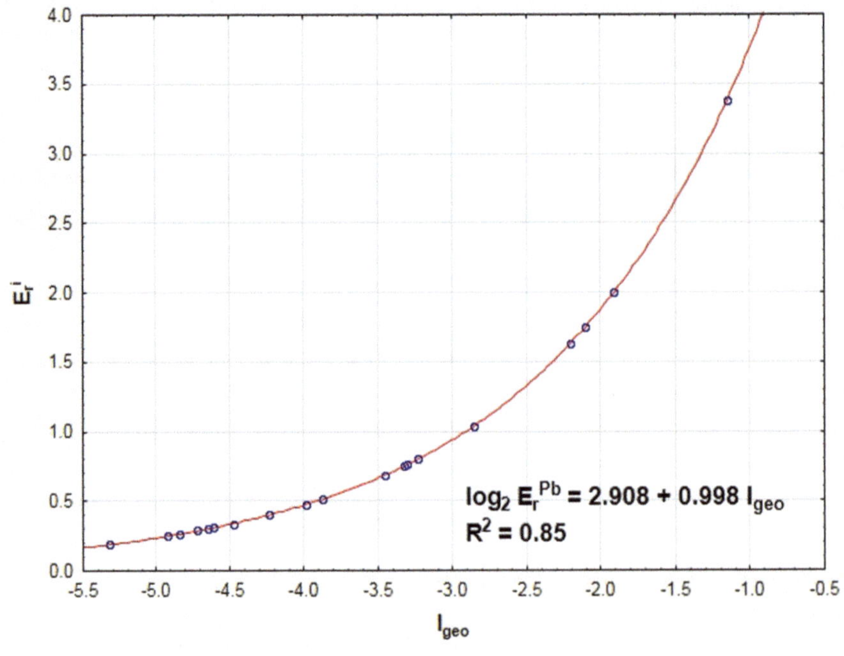

log2 Er^Pb = 2.908 + 0.998 Igeo
R2 = 0.85
Er^i
Igeo

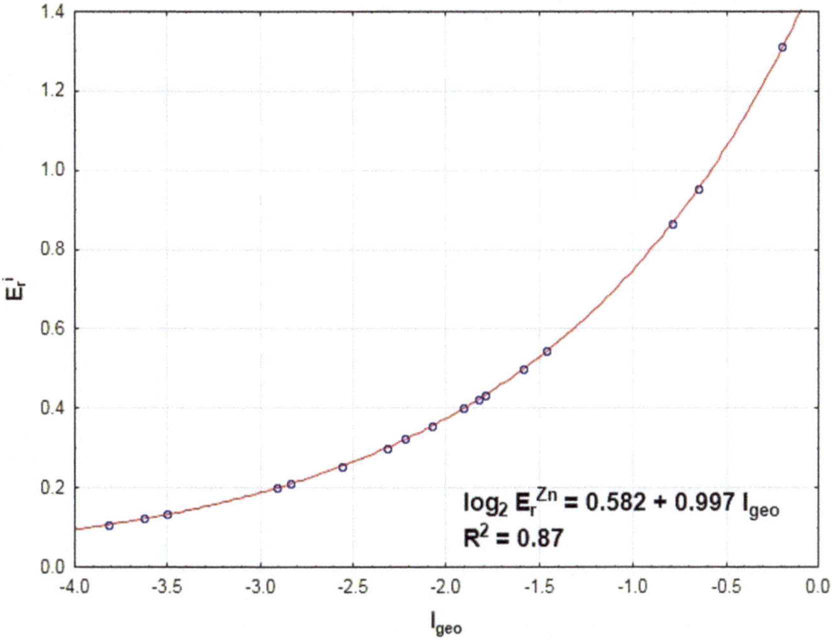

References

1. Wei, B., Yang, L.: A review of heavy metal contaminations in urban soils, urban road dusts and agricultural soils from China. Microchem. J. **94**(2), 99–107 (2010)
2. Chen, H., Chen, R., Teng, Y., Wu, J.: Contamination characteristics, ecological risk and source identification of trace metals in sediments of the Le'an River (China). Ecotoxicol. Environ. Saf. **172**, 85–92 (2016)
3. Islam, M.S., Ahmed, M.K., Raknuzzaman, M., Habibullah-Al-Mamun, M., Kundu, G.K.: Heavy metals in the industrial sludge and their ecological risk: a case study for a developing country. J. Geochem. Explor. **172**, 41–49 (2017). https://doi.org/10.1016/j.gexplo.2016.09.006
4. Raju, K.V., Somashekar, R., Prakash, K.: Heavy metal status of sediment in river Cauvery, Karnataka. Environ. Monit. Assess. **184**(1), 361–373 (2012)
5. Chakravarty, M., Patgiri, A.D.: Metal pollution assessment in sediments of the Dikrong river, NE India. J. Hum. Ecol. **27**(1), 63–67 (2009)
6. Atkinson, C.A., Jolley, D.F., Simpson, S.L.: Effect of overlying water pH, dissolved oxygen, salinity and sediment disturbances on metal release and sequestration from metal contaminated marine sediments. Chemosphere **69**(9), 1428–1437 (2007)
7. Zhong, A.-p., Guo, S.-h., Li, F.-m., Gang, L., Jiang, K.-x.: Impact of anions on the heavy metals release from marine sediments. J. Environ. Sci. **18**(6), 1216–1220 (2006)
8. Zhu, H.-n., Yuan, X.-z., Zeng, G.-m., Jiang, M., Liang, J., Zhang, C., Juan, Y., Huang, H.-j., Liu, Z.-f., Jiang, H.-w.: Ecological risk assessment of heavy metals in sediments of Xiawan port based on modified potential ecological risk index. Trans. Nonferrous Met. Soc. China **22**(6), 1470–1477 (2012)

9. Liu, J., Ma, K., Qu, L.: Ecological risk assessments and context-dependence analysis of heavy metal contamination in the sediments of mangrove swamp in Leizhou Peninsula, China. Mar. Pollut. Bull. **100**(1), 224–230 (2015)
10. Bocca, B., Alimonti, A., Petrucci, F., Violante, N., Sancesario, G., Forte, G., Senofonte, O.: Quantification of trace elements by sector field inductively coupled plasma mass spectrometry in urine, serum, blood and cerebrospinal fluid of patients with Parkinson's disease. Spectrochim. Acta Part B At. Spectrosc. **59**(4), 559–566 (2004)
11. Waisberg, M., Joseph, P., Hale, B., Beyersmann, D.: Molecular and cellular mechanisms of cadmium carcinogenesis. Toxicology **192**(2–3), 95–117 (2003)
12. Wang, W.-X.: Dietary toxicity of metals in aquatic animals: recent studies and perspectives. Chin. Sci. Bull. **58**(2), 203–213 (2013)
13. Ribeiro, D., Costa, S., Guilhermino, L.: A framework to assess the vulnerability of estuarine systems for use in ecological risk assessment. Ocean Coast. Manage. **119**, 267–277 (2016)
14. Hakanson, L.: An ecological risk index for aquatic pollution control. A sedimentological approach. Water Res. **14**(8), 975–1001 (1980)
15. Muller, G.: Index of geoaccumulation in sediments of the Rhine river. GeoJournal **2**, 108–118 (1969)
16. Perin, G., Craboledda, L., Lucchese, M., Cirillo, R., Dotta, L., Zanette, M., Orio, A.: Heavy metal speciation in the sediments of northern Adriatic Sea. A new approach for environmental toxicity determination. Heavy Met. Environ. **2**(1), 454–456 (1985)
17. Thurston, G.D., Spengler, J.D.: A multivariate assessment of meteorological influences on inhalable particle source impacts. J. Clim. Appl. Meteorol. **24**(11), 1245–1256 (1985)
18. Kulikowska, D., Gusiatin, Z.M., Bułkowska, K., Kierklo, K.: Humic substances from sewage sludge compost as washing agent effectively remove Cu and Cd from soil. Chemosphere **136**, 42–49 (2015)
19. Duodu, G.O., Goonetilleke, A., Ayoko, G.A.: Comparison of pollution indices for the assessment of heavy metal in Brisbane river sediment. Environ. Pollut. **219**, 1077–1091 (2016)
20. Dahms, S., Baker, N., Greenfield, R.: Ecological risk assessment of trace elements in sediment: a case study from Limpopo, South Africa. Ecotoxicol. Environ. Saf. **135**, 106–114 (2017)
21. Zhang, P., Qin, C., Hong, X., Kang, G., Qin, M., Yang, D., Pang, B., Li, Y., He, J., Dick, R.P.: Risk assessment and source analysis of soil heavy metal pollution from lower reaches of Yellow river irrigation in China. Sci. Total Environ. **633**, 1136–1147 (2018)
22. Liu, J.-J., Ni, Z.-X., Diao, Z.-H., Hu, Y.-X., Xu, X.-R.: Contamination level, chemical fraction and ecological risk of heavy metals in sediments from Daya Bay, South China Sea. Mar. Pollut. Bull. **128**, 132–139 (2018)
23. Li, R., Li, R., Chai, M., Shen, X., Xu, H., Qiu, G.: Heavy metal contamination and ecological risk in Futian mangrove forest sediment in Shenzhen Bay, South China. Mar. Pollut. Bull. **101**(1), 448–456 (2015)
24. Duodu, G.O., Goonetilleke, A., Ayoko, G.A.: Potential bioavailability assessment, source apportionment and ecological risk of heavy metals in the sediment of Brisbane river estuary, Australia. Mar. Pollut. Bull. **117**(1–2), 523–531 (2017)
25. Sofianska, E., Michailidis, K.: Assessment of heavy metals contamination and potential ecological risk in soils affected by a former Mn mining activity, drama district, northern Greece. Soil Sedim. Contam. Int. J. **25**(3), 296–312 (2016)
26. Mohseni-Bandpei, A., Ashrafi, S.D., Kamani, H., Paseban, A.: Contamination and ecological risk assessment of heavy metals in surface soils of Esfarayen city, Iran. Health Scope **6**(2) (2017)
27. Raj, D., Chowdhury, A., Maiti, S.K.: Ecological risk assessment of mercury and other heavy metals in soils of coal mining area: a case study from the eastern part of a Jharia coal field, India. Hum. Ecol. Risk Assess. Int. J. **23**(4), 767–787 (2017)
28. Hu, B., Zhou, J., Liu, L., Meng, W., Wang, Z.: Assessment of heavy metal pollution and potential ecological risk in soils of Tianjin sewage irrigation region, North China. J. Environ. Anal. Toxicol. **7**(425), (2017). https://doi.org/10.4172/2161-0525.1000425

29. Al-Rawi, A.: Assessment of Heavy Metals in the Sediments of the Euphrates River in Al-Anbar Province. University of Anbar (2012)
30. Mortatti, J., de Moraes, G.M., Probst, J.-L.: Heavy metal distribution in recent sediments along the Tietê river basin (São Pauro, Brazil). Geochem. J. **46**(1), 13–19 (2012)
31. Salah, E.A., Turki, A.M., Mahal, S.N.: Chemometric evaluation of the heavy metals in urban soil of Fallujah city, Iraq. J. Environ. Prot. **6**(11), 1279 (2015)
32. Bortey-Sam, N., Nakayama, S., Akoto, O., Ikenaka, Y., Baidoo, E., Mizukawa, H., Ishizuka, M.: Ecological risk of heavy metals and a metalloid in agricultural soils in Tarkwa, Ghana. Int. J. Environ. Res. Public Health **12**(9), 11448–11465 (2015)
33. USEPA: US Environmental Protection Agency: Screening Level Ecological Risk Assessment Protocol for Hazardous Waste Combustion Facilities, vol. 3 (1999)
34. Protano, C., Zinnà, L., Giampaoli, S., Spica, V.R., Chiavarini, S., Vitali, M.: Heavy metal pollution and potential ecological risks in rivers: a case study from southern Italy. Bull. Environ. Contam. Toxicol. **92**(1), 75–80 (2014)
35. Gupta, S.K., Chabukdhara, M., Kumar, P., Singh, J., Bux, F.: Evaluation of ecological risk of metal contamination in river Gomti, India: a biomonitoring approach. Ecotoxicol. Environ. Saf. **110**, 49–55 (2014)
36. Wang, N., Wang, A., Kong, L., He, M.: Calculation and application of Sb toxicity coefficient for potential ecological risk assessment. Sci. Total Environ. **610**, 167–174 (2018)
37. Paterson, G., Macken, A., Thomas, K.V.: The need for standardized methods and environmental monitoring programs for anthropogenic nanoparticles. Anal. Methods **3**(7), 1461–1467 (2011)
38. Zhao, S., Feng, C., Yang, Y., Niu, J., Shen, Z.: Risk assessment of sedimentary metals in the Yangtze Estuary: new evidence of the relationships between two typical index methods. J. Hazard. Mater. **241**, 164–172 (2012)

[illegible]

Potential Ecological Risk Assessment of Heavy Metals in Iraqi Soils: Case Studies

Emad Al-Heety and Wahran Saod

Abstract The aim of this study is to estimate the potential ecological risk of heavy metals in soils in different cities, Iraq. The used data was obtained from previous published studies. The index of ecological risk (ErI) and index of the overall potential ecological risk (RI) were calculated employing Hakanson's methodology. Generally, the obtained results showed that Cd was the major contributor to the potential risk in soils of Baghdad, Basrah, Duhok, Erbil, Fallujah, and Kirkuk cities. In soils of Al-Nasiriya, Babylon and Hawega cities, the major contributor to the potential risk is Ni element. The spatial distribution of the RI showed that heavy risk was recorded in Baghdad city soil and moderate risk in soils of Basrah, Duhok, Erbil and Kirkuk while the risk in soils of the other cities was light. The spatial distribution reflects impact of the anthropogenic activities as a source of the metal pollution of the soils. The results showed that the factor controlling the estimation of E_r^i was the reference level value of the metal and using different reference values leads to overestimation or underestimation of the E_r and in turn RI. This work represents the first attempt to assess the ecological risk of heavy metal in Iraqi soils except Baghdad's soil.

Keywords Heavy metal · Pollution · Soil · Ecological risk · Iraq

1 Introduction

Because of toxicity, bioaccumulation, abundance and persistence of the heavy metals, the heavy metals contamination has attracted global concern as serious environmental problem [1–3]. Recently, soil heavy metal pollution has been paid significant attention as worldwide issue. Mainly, accumulations of the heavy metals in soils

E. Al-Heety (✉)
Department of Applied Geology, University of Anbar, Ramadi, Iraq
e-mail: emadsalah@uoanbar.edu.iq

W. Saod
Department of Chemistry, University of Anbar, Ramadi, Iraq
e-mail: sc.wahran.s@uoanbar.edu.iq

Y. T. Mustafa et al. (eds.), *Recent Researches in Earth and Environmental Sciences*,
Springer Proceedings in Earth and Environmental Sciences,
https://doi.org/10.1007/978-3-030-18641-8_6

and especially in urban soil come from industrial activities, coal and fuel burning, vehicles emissions, mining operations, fertilizers and pesticides use, municipal solid waste disposal, and other wastes [1, 4]. Heavy metals in soils spread and accumulate in plants and consume by human. The heavy metals cumulate in greasy tissues and influence the actions of nerves, endocrine, and immune systems, normal cellular metabolism, etc. [5–7]. The ecological risk assessment is tool to some extent specifies the probability of an adverse impact on an organism or ecosystem due to exposure to environmental stressors, such as, chemical or biological pollution. Different methods were developed to calculate the ecological risk of heavy metal, like potential ecological risk index (E_r^i) [8], index of geoaccumulation (Igeo) [9], and modified potential risk index (M_r^i) [10]. Index of geoaccumulation (Igeo) and potential ecological risk index (E_r^i) widely employed as a quantitative measure of the potential risk of heavy metals in soils. These two indices (Igeo and E_r^i) have been extensively used to assess E_r^i of heavy metals in polluted soils by heavy metals [11–14]. There are few studies on the assessment of the ecological risk of heavy metals in Iraqi soils [15, 16]. The primary aims of this study are: (1) to estimate index of potential ecological risk (E_r^i) and risk index (RI) of heavy metals in different cities in Iraq using the available data; (2) to assess the spatial variation of the E_r^i and RI in Iraq; (3) to identify the heavy metal that contributes more than the other metals to ecological risk in each city and across Iraq.

2 Materials and Methodology

The case studies were selected depending on the available data of heavy metal level in soil. The used data in this work was listed in Table 1. The E_r^i and RI of soil heavy metal in the study area are estimated using Hakanson's methodology [8]. The E_r^i is estimated by the following relations:

$$E_r^i = T_f^i \times C_f^i \tag{1}$$

$$C_{f^i} = \frac{C_{s^i}}{C_{r^i}} \tag{2}$$

$$E_r^i = T_{f^i} \times \frac{C_{s^i}}{C_{r^i}} \tag{3}$$

where E_r^i is the potential ecological risk factor of metal i, T_f^i is the toxic response factor of metal i. The T_f values of heavy metals are 30, 2, 5, 5, 5, and 1 for Cd, Cr, Cu, Ni, Pb and Zn, respectively. C_f^i is the metal i concentration in an sediment or soil sample, and C_r^i is the reference value of metal i. The RI is a sum of the E_r^i for all heavy elements at the sampling sites:

$$RI = \sum_{i}^{n} E_r^i \tag{4}$$

The categories of E_r^i and *RI* as classified by Hakanson [8] are: light potential ecological risk ($E_r^i < 40$), moderate potential ecological risk ($40 \leq E_r^i < 80$), heavy potential ecological risk ($80 \leq E_r^i < 160$), severe potential ecological risk ($160 \leq E_r^i < 320$), and very severe ecological risk ($320 \leq E_r^i$). $RI < 150$, light potential ecological risk; $150 \leq RI < 300$, moderate potential ecological risk; $300 \leq RI < 600$, heavy potential; $RI > 600$, severe potential. We used the data listed in Table 1 to calculate the E_r^i and *RI* of some heavy metals in the studied cases.

3 Results and Discussion

Results of E_r^i and *RI* in the studied cases are shown in Figs. 1 and 2, respectively. The E_r^i values of heavy metals in soils of the studied cases were in descending order: $Ni > Cr > Cu > Pb > Zn$ (Al-Nasiriya), $Ni > Pb > Cr$ (Babylon), $Cd > Ni > Cu > Pb > Zn$ (Baghdad), $Cd > Pb > Ni > Cr$ (Basrah), $Cd > Ni > Pb > Zn$ (Duhok), $Cd > Cu > Pb > Cr > Zn$ (Erbil), $Cd > Ni > Cr > Cu > Pb > Zn$ (Fallujah), $Ni > Cr > Cu > Zn$ (Hawega) and $Cd > Cu > Pb > Ni > Cr > Zn$ (Kirkuk). As shown in Fig. 1, except for Ni in Al-Nasiriya and Hawega Cities, the E_r^i values of Cr, Cu, Ni, Pb, and Zn indicate light potential ecological risk to soils of the other cities. The E_r^i of Ni in soils of Al-Nasiriya and Hawega Cites showed moderate risk. The E_r^i values of Cd were light in Erbil and Fallujah, heavy in Basrah and Kirkuk, and severe in Baghdad and Duhok. In the light of the *RI* value, metal pollution in soils of the studied cities can be graded in descending order as follows: Baghdad > Kirkuk > Erbil > Duhok > Basrah > Haweja > Al-Nasiriya >

Table 1 The concentrations of heavy metals in soils of the study area

Case	Concentration of metal (mg/kg)						References
	Cd	Cr	Cu	Ni	Pb	Zn	
Al-Nasiriya		116.3	27.3	147.1	16.84	57.3	[17]
Babylon		33.4		98.84	33.53		[18]
Baghdad	5.52		91.9	111.4	153.7	133.3	[19]
Basrah	2.80	60.90		36.90	115.00		[20]
Duhok	3.42			78.73	88.28	308.86	[21]
Erbil	3.35	37.5	128		181.5	5.48	[22]
Fallujah	0.64	11.59	201	8.96	3.82	5.50	[23]
Hawega		310.5	35.7	152.3		51.33	[24]
Kirkuk	3.03	27.56	103.1	161.93	232.79	14.33	[25]
Reference values	0.6	25	16	16	40	90	[26]

Babylon > Fallujah. Three grades of *RI* were reported in the studied cases, light, moderate and heavy risk. Cd is the main contributor in the *RI* of Baghdad, Duhok, Erbil, Kirkuk, Basrah, and Fallujah. In Babylon, Al-Nasiriya, Hawega, and Erbil, the main contributor in *RI* was Ni. In general, the contribution of the single metal in *RI* was observed in the following order: *Cd* > *Ni* > *Cu* > *Cr* > *Pb* > Zn. According to Hakanson's methodology, E_r^i value depends on concentration of the metal in the soil sample, reference concentration and coefficient the toxicity response. Regardless of metal concentration and its toxicity coefficient, the E_r^i was estimated for four metals using three different reference values [26–28]. Using different reference metal values could cause an overestimation or underestimation of the E_r^i. For example, the world average data of Cd has been given overestimated value of E_r^i, while the Canadian guidelines recorded underestimation of the E_r^i. The lack of abundance of updated reference metal concentrations could lead to an overestimation or underestimation of the E_r^i [29]. The accurate assessment of the E_r^i requires routine upgrading of the reference concentrations periodically at the regional scales in each country, especially in geological regions containing sensitive ecological habitats [30]. The *RI* value is related with the kind and amount of the heavy metal. A high *RI* value reflects a high concentration of the metal and high toxicity, while a lesser *RI* value is a lesser level of the metal and a slighter toxicity [31]. The obtained results were compared with those available for soils in Baghdad City only because of the lack of published studies about the assessment of ecological risk of heavy metals in the soils of the cities. Al-Anbari et al. [15] investigated the ecological risk assessment of heavy metals in the urban soil in Al-Karrada district, Baghdad affected by various anthropogenic activities.

They recorded very high single ecological risk E_r^i of Cd (672) and high overall potential ecological risk index *RI* (720) in the study area. The difference between the E_r^i of Cd (262.5) obtained in the current study and that obtained in study [15], can be attributed to difference in the selected reference concentration of Cd although the concentration of Cd in the present study was more than that in study [15]. They used 0.2 as reference concentration of Cd while in the current study was 0.6. Mohammed and Abdullah [16] estimated the ecological risk E_r^i of heavy metals in soil of the northern site of East Baghdad oil field. The E_r^i values were classified as low ecological risk for the measured heavy metals including Cd. The high concentration of Cd in soils Baghdad, Duhok, Kirkuk and Basrah is attributed to the human activities, such as high traffic density, the industrial activity, use of the phosphate fertilizer and the atmospheric pollution [15, 19–21, 25]. The spatial distribution of the *RI* (Fig. 2) showed heavy ecological risk in soil of Baghdad City which can be attributed to the extensive various human activities because Baghdad is the largest residential, industrial and commercial center in Iraq. The moderate *RI* in Basrah and Kirkuk may be interpreted in terms of both are centers of urbanization, industrial and oil production. The moderate *RI* in soil of Duhok City also comes from human activities. The light *RI* in the other cities may reflect less anthropogenic activities compared with those in Baghdad, Basrah, Duhok, Erbil and Kirkuk. The spatial distribution shows large impact of the anthropogenic origin of metal contamination in Iraqi soils rather than the geogenic origin.

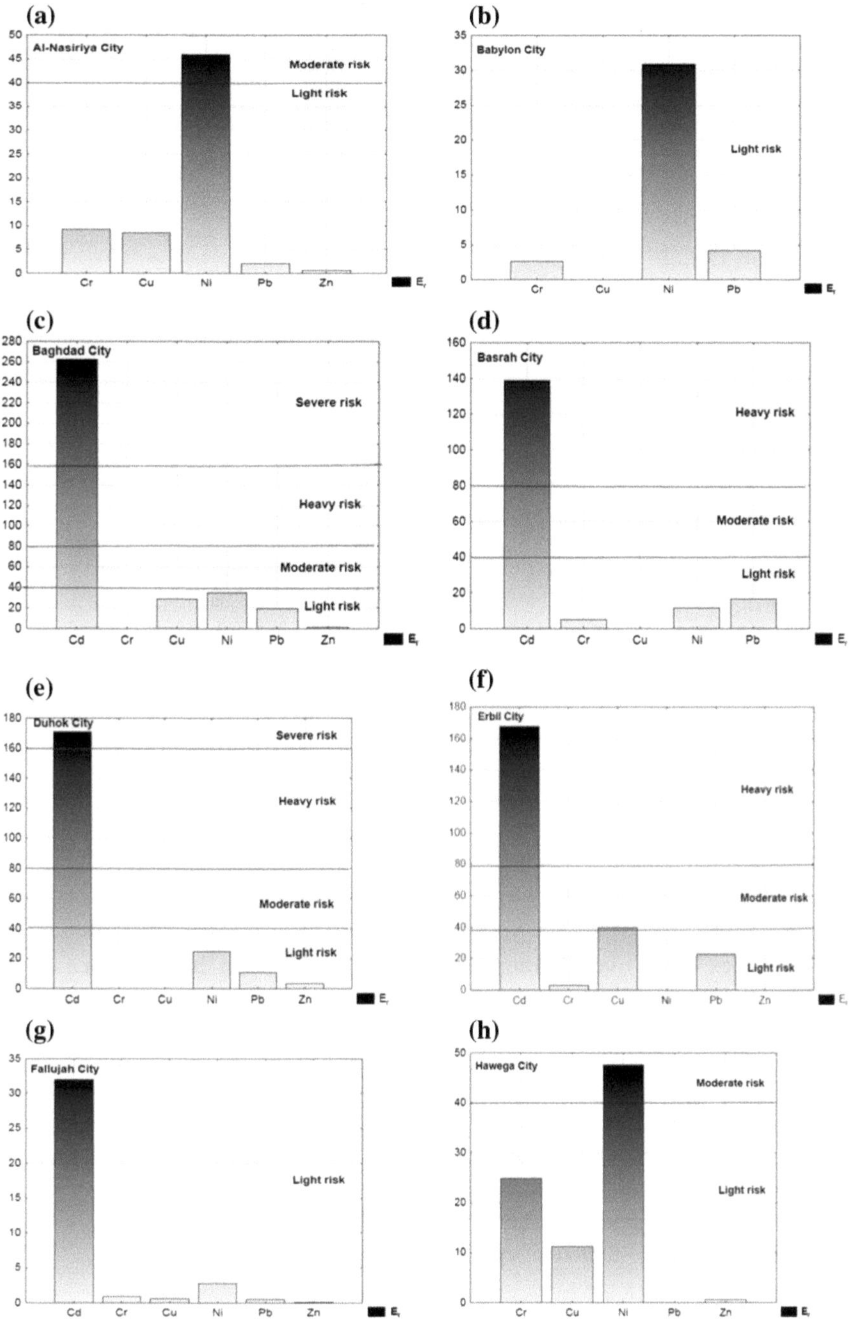

Fig. 1 Ecological risk factor (E_r^i) of heavy metals in soils of the study area

Fig. 1 (continued)

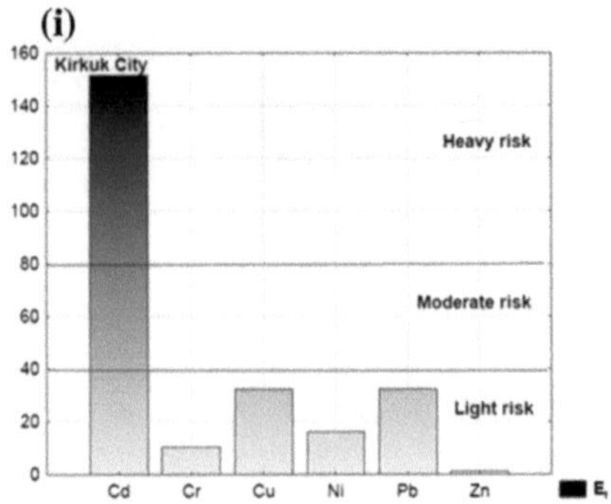

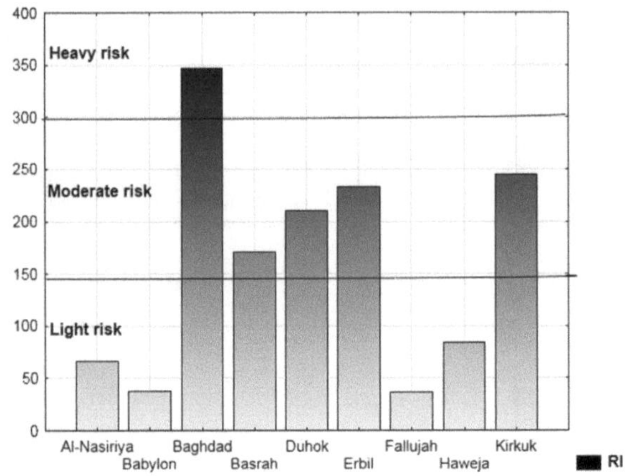

Fig. 2 Spatial variation of the potential ecological risk index (RI) of heavy metals in soils of the study area

4 Conclusions

1. The index of the overall potential ecological risk (RI) in the study area ranges from light to heavy.
2. Cd showed a severe ecological risk in soil of Baghdad City and moderate ecological risk in Duhok, Erbil, Kirkuk and Basrah soils.
3. The factor controlled calculation of the ecological risk is the reference concentration of metal.
4. Using non updating reference concentration value could cause overestimation or underestimation of ecological risk.
5. The spatial distribution of the *RI* reflects the anthropogenic source for soil contamination in the studied cases with heavy metals.

References

1. Wei, B., Yang, L.: A review of heavy metal contaminations in urban soils, urban road dusts and agricultural soils from China. Microchem. J. **94**(2), 99–107 (2010)
2. Chen, H., Chen, R., Teng, Y., Wu, J.: Contamination characteristics, ecological risk and source identification of trace metals in sediments of the Le'an river (China). Ecotoxicol. Environ. Saf. **125**, 85–92 (2016)
3. Islam, M.S., Ahmed, M.K., Raknuzzaman, M., Habibullah-Al-Mamun, M., Kundu, G.K.: Heavy metals in the industrial sludge and their ecological risk: a case study for a developing country. J. Geochem. Explor. **172**, 41–49 (2017)
4. Liu, J., Ma, K., Qu, L.: Ecological risk assessments and context-dependence analysis of heavy metal contamination in the sediments of mangrove swamp in Leizhou Peninsula, China. Mar. Pollut. Bull. **100**(1), 224–230 (2015)
5. Waisberg, M., Joseph, P., Hale, B., Beyersmann, D.: Molecular and cellular mechanisms of cadmium carcinogenesis. Toxicology **192**(2–3), 95–117 (2003)
6. Bocca, B., Alimonti, A., Petrucci, F., Violante, N., Sancesario, G., Forte, G., Senofonte, O.: Quantification of trace elements by sector field inductively coupled plasma mass spectrometry in urine, serum, blood and cerebrospinal fluid of patients with Parkinson's disease. Spectrochim. Acta Part B **59**(4), 559–566 (2004)
7. Wang, W.-X.: Dietary toxicity of metals in aquatic animals: recent studies and perspectives. Chin. Sci. Bull. **58**(2), 203–213 (2013)
8. Hakanson, L.: An ecological risk index for aquatic pollution control. A sedimentological approach. Water Res. **14**(8), 975–1001 (1980)
9. Muller, G.: Index of geoaccumulation in sediments of the Rhine river. GeoJournal **2**, 108–118 (1969)
10. Zhu, H.-n., Yuan, X.-z., Zeng, G.-m., Jiang, M., Liang, J., Zhang, C., Juan, Y., Huang, H.-j., Liu, Z.-f., Jiang, H.-w.: Ecological risk assessment of heavy metals in sediments of Xiawan port based on modified potential ecological risk index. Trans. Nonferrous Met. Soc. China **22**(6), 1470–1477 (2012)
11. Jiao, X., Teng, Y., Zhan, Y., Wu, J., Lin, X.: Soil heavy metal pollution and risk assessment in Shenyang industrial district, Northeast China. PLoS One **10**(5), e0127736 (2015)
12. Mohseni-Bandpei, A., Ashrafi, S.D., Kamani, H., Paseban, A.: Contamination and ecological risk assessment of heavy metals in surface soils of Esfarayen city, Iran. Health Scope **6**(2) (2017)
13. Giri, S., Singh, A.K.: Ecological and human health risk assessment of agricultural soils based on heavy metals in mining areas of Singhbhum copper belt, India. Hum. Ecol. Risk Assess. Int. J. **23**(5), 1008–1027 (2017)
14. Raj, D., Chowdhury, A., Maiti, S.K.: Ecological risk assessment of mercury and other heavy metals in soils of coal mining area: a case study from the eastern part of a Jharia coal field, India. Hum. Ecol. Risk Assess. Int. J. **23**(4), 767–787 (2017)
15. Al-Anbari, R., Al Obaidy, A.H.M., Ali, F.H.A.: Manuscript info abstract. Int. J. **3**(2), 104–110 (2015)
16. Mohammed, M.S., Abdullah, E.J.: Heavy metals pollution assessment of the soil in the northern site of east Baghdad oil field, Iraq. Iraqi J. Sci. **57**(1A), 175–183 (2016)
17. Benni, T.J.: Distribution of some heavy metals in the recent sediments of Al-Nasiriya area, mesopotamia plain, south Iraq. Iraqi Bull. Geol. Min. **10**(1), 73–92 (2014)
18. Manii, J.K.: Using GIS to study the probability pollution of surface soil in Babylon province, Iraq. J. Appl. Geol. Geophys. **2**, 14–18 (2014)
19. Sultan, M.A.: Evaluation of soil pollution by heavy metals in Baghdad city using GIS. In: The 1st International Applied Geological Congress, Department of Geology, Islamic Azad University-Mashad Branch, Iran, pp. 26–28 (2010)
20. Abass, M.H., Hassan, Z.K., Al-Jabary, K.M.: Assessment of heavy metals pollution in soil and date palm (*Phoenix dactylifera* L.) leaves sampled from Basra/Iraq governorate. Adv. Environ. Sci. **7**(1), 52–59 (2015)

21. Yousif, K.M.: Preliminary assessment of heavy metal in selected sites within Duhok city, Iraq. Am. Sci. Res. J. Eng. Technol. Sci. (ASRJETS) **26**(2), 316–325 (2016)
22. Sirwan, S., Ali, I., Mzgin, A., Karwan, A.: The study of contaminated toxic heavy metals of traffic street dust and soils in Erbil City. In: Proceedings of ICEEE Conference Ist International Conference on Ecology, Environment, and Energy, June 2014, Ishik University, Erbil (2014)
23. Salah, E.A., Turki, A.M., Mahal, S.N.: Chemometric evaluation of the heavy metals in urban soil of Fallujah city, Iraq. J. Environ. Prot. **6**(11), 1279 (2015)
24. Ali, H.A.: Heavy metals concentrations in surface soils of the Haweja area south western of Kirkuk, Iraq. Kirkuk Univ. J. Sci. Stud. **2**(3), 35–48 (2007)
25. AL-Jumaily, H.A.: Geochemical evaluation of heavy metals pollution of industrial quarter soils at Kirkuk city. Northern Iraq. Kirkuk Univ. J. Sci. Stud. **4**(1), 1–11 (2009)
26. SW-846 Online: Waste Hazards. Waste—Test Methods. http://www.epa.gov/osw/hazard/testmethods/sw846/online/index.htm (2012)
27. Bowen, H.J.M.: Environmental Chemistry of the Elements. Academic Press (1979)
28. CEPA, C.E.P.A.: Canadian Soil Quality Guidelines for the Protection of Environmental and Human Health (2007)
29. Protano, C., Zinnà, L., Giampaoli, S., Spica, V.R., Chiavarini, S., Vitali, M.: Heavy metal pollution and potential ecological risks in rivers: a case study from southern Italy. Bull. Environ. Contam. Toxicol. **92**(1), 75–80 (2014)
30. Gupta, S.K., Chabukdhara, M., Kumar, P., Singh, J., Bux, F.: Evaluation of ecological risk of metal contamination in river Gomti, India: a biomonitoring approach. Ecotoxicol. Environ. Saf. **110**, 49–55 (2014)
31. Zhang, P., Qin, C., Hong, X., Kang, G., Qin, M., Yang, D., Pang, B., Li, Y., He, J., Dick, R.P.: Risk assessment and source analysis of soil heavy metal pollution from lower reaches of Yellow river irrigation in China. Sci. Total Environ. **633**, 1136–1147 (2018)

Using a Mix of Three Microbial Strains on Fermentation and Aerobic Stability of Grass Silage

Vahel Jaladet Taha

Abstract This study was conducted to investigate the effect of using a mix of *Lactobacillus (L.) plantarum, L. casei* and *Saccharomyces cerevisiae* as silage inoculate to meadow grass on silage fermentation characteristics, silage quality and aerobic stability. A sample of Meadow grass (MG) was collected directly from the forage harvester during a first cut harvest. The fresh, chopped MG sample was divided into 2 equal portions and either treated with the test additive (EM) or received an equal volume of water (control). The dosage of EM Silage was based on 80 ml per ton product. Samples were analysed statistically as randomise complete (1 × 2) design using an ANOVA procedure of Genstat 15. Ensiled sample was found to have lower dry matter (DM), crude protein (CP), water soluble carbohydrate (WSC) and yeasts count, and higher mould count compared fresh samples. Additionally, treating MG with EM increased concentration of total volatile fatty acids (tVFA), acetic and propionic acids and total alcohol by approximately (25, 27, 47, 300% respectively) and reduced lactic acid by approximately 13% compared to control samples. Aerobic stability results showed that control MG treatment reached a temperature 3 °C above ambient in ~6.8 days whereas the EM treated MG took approximately 14 days get to the same temperature after exposing to air. Therefore, using this mix of inoculate would enhance silage quality and protect silage from aerobic deterioration.

Keywords Silage additive · Inoculate · Lactic acid bacteria

1 Introduction

Silage defined as a highly fermented water-soluble carbohydrate in the forage to lactic and butyric acid resulting reduction of pH and used for animal feeding when fresh forage is scarce [1]. In Europe, silage industry increased by approx. 160 million ton

V. J. Taha (✉)
Department of Animal Production, College of Agriculture, University of Duhok, Duhok, Kurdistan Region, Iraq
e-mail: vahel.taha@uod.ac

Y. T. Mustafa et al. (eds.), *Recent Researches in Earth and Environmental Sciences*, Springer Proceedings in Earth and Environmental Sciences,
https://doi.org/10.1007/978-3-030-18641-8_7

per year of forage dray matter since the mid of the last century [1]. Nowadays, silage may represent more than 80% of winter forages in ruminant diets in the UK [2]. The ensiling process could be divided into three different steps: initial aerobic step, fermentation step and second aerobic step [3]. Silage fermentation was undertaken by different species of microflora associated silage herbage and was largely left to chance to represent an uncontrolled process [4]. However, using different silage additives and/or physical approaches (wilting) plus clamp management would enhance fermentation process resulting an improvement in silage quality [1]. This aim could be achieved by using silage additive especially silage inoculate [5]. Silage inoculants have been developed for their ability to promote a beneficial fermentation that maximizes the nutritive value of the silage for ruminant animals [1]. Since the beginning of the last century, silage inoculations have been successfully used in France and other European countries [6]. Supplemented inoculated bacteria to forage at ensiling would cause a rapid reduction of the silage pH and prevent the growth of undesirable microorganisms such as clostridia [7]. In addition, supplemented inoculated bacteria at ensiling would reduce DM losses during the fermentation period [7]. Hence, the used of inoculated bacteria as a silage additive has been widely increased [6].

Lactic acid bacteria (LAB) are the common commercial silage additives used globally [8]. Lactic acid bacteria such as *Lactobacillus (L.) buchneri*, *L. plantarum* and *L. casei* were developed as silage inoculant to enhance silage fermentation increasing acetic acid concentration and improve the aerobic stability of the silage [9]. Merry and Davies [1] outlined that LAB could be divided into two groups: homo-fermentative and hetro-fermentative LAB. Homo-fermentative LAB contained one or more strains species: *L. plantarum, L. lactis, Pediococcus* spp. and *Enterococcus* spp. The main final product of these species is lactic acid. Hetro-fermentative LAB such as *L. buchneri*, which produce lactic acid, acetic acid, ethanol and carbon dioxide as a final product [10]. Homo-fermentative LAB caused a rapid decrease in silage herbage pH and reduced losses of DM to a minimum level (2–3%) [11]. Moreover, these strains of LAB are the traditional inoculant bacteria used in the market [12]. In addtion, the fast decrease in pH prevents the growth of many undesirable microorganisms inside the silo such as clostridia, which are responsible for producing butyric acid. Mixing different strain of silage inoculate may act synergistically with each other and improved silage quality and aerobic deterioration [13]. Yeasts such as *Saccharomyces* spp. are used in food science to improve fermentation such as yoghurt making [9]. Several studies [9, 11, 12] showed that using *Saccharomyces* spp. might improve feed conversion ratio, reduce ruminal acidosis, and mitigating methane emissions. A lack of studies for using yeasts especially *Saccharomyces* spp. mixed with LAB as silage inoculate for enhancing silage quality either ensiling or after air exposure. Therefore, the objective of this study was to investigate the effect of using a mix of *L. plantarum, L. casei* and *Saccharomyces cerevisiae* as silage inoculate to meadow grass on silage fermentation characteristics, silage quality and aerobic stability.

2 Materials and Methods

A sample of meadow grass (MG) was collected directly from the forage harvester during a first cut harvest on 1st May 2017. The fresh, chopped MG sample was divided into 2 equal portions and either treated with the test additive (EM treated; batch No. 1509010001) or received an equal volume of water. The dosage of EM silage was based on 80 ml per ton product. Product microbial composition was *Lactobacillus casie* 137 × 108 bacteria/ml, *Lactobacillus plantarum* 865 × 104 bacteria/ml and *Saccharomyces cerevisiae* 26 × 105 bacteria/ml. The inoculate treatment was prepared by dissolving 2 g of a mix of freeze-dried inoculated bacteria and 30 g of table sugar (as media) in 1000 ml deionised water and incubate for 16 h at 30 °C. Mini silos were created in six replications for each of the two experimental treatments. During the filling of the mini-silos, subsamples (fresh material) were collected for each min-silo. Fresh material subsamples were vacuum packed and stored at −20 °C for proximate analysis. Each experimental silo was lined with a plastic bag and filled with approximately 1.5 kg from either treated or control MG and consolidated well. The neck of the plastic bag liner was then sealed using silage tape and approximately 0.5 kg of sand was placed on the top of each mini-silo. The weight of each mini-silos before and immediately after filling with MG and sealing were recorded. Following a 100-day ensiling period the mini silos were weighed (in order to calculate DM loss), opened and silage samples were taken. Ensiled samples were vacuum packed and stored at −20 °C for proximate analysis and fermentation characteristics.

All frozen ensiled samples for both treatments were defrosted at 4 °C for 16 h and analysed for proximate analysis and fermentation characteristics. The ammonia nitrogen (NH_3-N) concentration was measured according to MAFF [13] using a Buchi AutoKjeldahl Unit K-370 (buchi labortechnik AG CH-9.0, Flawil, Switzerland), and pH determined by the method of MAFF [13] using a pH meter (Jenway, Stone, Staffordshire) with calibration using pH 4 and pH 7 buffers. Water-soluble carbohydrate (WSC) concentration was determined by spectrophotometrically (at wavelength 620 nm) as the blue-green complex which is formed when carbohydrate is heated with Anthrone in sulphuric acid using MAFF [13]. Defrosted samples of both fresh and ensilage forage from each mini-silo were oven dried at 105 °C overnight according to method number 935.29 [14]. Dried samples were milled through a 1 mm screen and used to determine ash by using 1 g of dried sample in a muffle furnace at 550 °C for 5 h and crude protein (CP) using a C/N analyser [type FP-528, LECO Instruments, St. Joseph, MI, USA] operating the Dumas method number 990.03 [14]. Samples of fresh and ensiled grass were vacuum packed and sent to an approved laboratory [Sciantec, UK or Northern Hygiene, UK for Lactic Acid Bacteria counts] to analysis manually for mould, yeast and lactobacilli counting according to method published by Borreani and Tabacoo [15]. Silage samples were analyzed additionally for silage characteristics including lactic acids (using HPLC machine according to method published by Canale et al. [16]) total and individual volatile fatty acid (VFA) (using gas chromatography machine according to method published by Cruwys et al. [17]) and total and individual alcohol concentration (using

HPLC according to method published by Borreani and Tabacoo [15]) at the same laboratory.

Aerobic stability of the ensiled material was determined by temperature difference to ambient. Approximately 500 g of ensiled MG samples from each of 12 experimental mini-silos were placed individually in insulated boxes open to the atmosphere at 18–20 °C for 20 days. One themochron iButtons [iButton THERMO-S-KIT Thermochron Starter Kit] was stuck in each box and another one in the silos storage room to record temperature every 10 min. The temperature of ambient and each box were observed continuously during the aerobic stability period (20 days).

All obtained data were analysed statistically using an ANOVA procedure of Genstat (GenStat version 15, VSN International Ltd., UK). Proximate analysis of silage samples was analysed using randomise complete (1×2) design.

3 Results and Discussion

3.1 Chemical Analysis

The chemical analysis results of both fresh and ensiled forage are shown in Table 1. Target DM for the crop was achieved with a mean DM for the fresh material of 370.8 g/kg. The enisled material was found to have a lower DM value (341.6 g/kg) compared to fresh forage (370.8 g/kg) which probably due to evaporative and volatile losses during the ensiling phase. The silages were kept in a plastic bag inside mini-silos, and the integrity of the bags would inhibit effluent escape. Mayne and Gordon [18] suggested that DM losses could occur in three ways: effluent escape, surface waste and invisible losses. The invisible losses could include fermentation characteristic and plant respiration during the fermentation period especially the initial fermentation phase [18]. Therefore, the DM losses in the present study could be due to an invisible loss according to Mayne and Gordon [18]. In addition, silage DM losses could be affected by the activity of some microorganisms in the silage clamp such as Enterobacteria [5]. A similar result was found by Wambacq et al. [19] who reported that ensiled mix of alfalfa and ryegrass had a lower DM compared to fresh forage. They went to show that the losses of DM bases could be due to the effect of microbial activity inside silage silos during the fermentation period. Similarly, Taha [5] also found that ensiled bean, pea and grass reduced DM compared to fresh forages by approximately 18, 20 and 14% for bean, pea and grass respectively. However, in the current study treated MG forage with test additive EM had no effect on DM of ensiled forage compared to untreated silage samples. Reich and Kung [20] highlighted that one of the disadvantages of using LAB, as a silage additive would be their effect on increasing silage DM losses at fermentation process.

Forage CP was found to be low, but this can only be attributed to the nature of the material harvested compared to other finding. Ensiling the MG significantly reduced

Table 1 Effect of using three microbial strains as a silage additive on the chemical analysis of fresh and ensiled meadow grass

Parameter	Fresh MG	Ensiled MG	Control	EM treated	Treatment SED	Interaction SED	Ensiling *P*-value	Treatment *P*-value	Interaction *P*-value
DM g/kg	370.8	341.6	356.9	355	5.27	7.46	<0.001	0.79	0.37
OM g/kg DM	967.1	965.2	966.6	965.1	1.15	1.87	0.75	0.66	0.48
CP g/kg DM	72.75	68.58	71.08	70.25	1.902	2.691	0.041	0.66	0.49
EE g/kg DM	11.1	10.9	11.3	10.7	0.38	0.48	0.36	0.37	0.25
WSC g/kg DM	72.9	12.1	41.3	43.7	2.19	3.09	<0.001	0.29	0.8
Mould[a]	1.88	3.12	2.65	2.35	0.315	0.445	<0.001	0.35	0.19
Yeast[a]	3.18	2.45	2.93	2.7	0.309	0.437	0.028	0.47	0.25
Lactobacilli[a]			1.645	2.45	0.144			<0.001	

EM experimental treatment, *ME* meadow grass, *SED* standard error of division, *P* probability, *DM* dry matter, *OM* organic matter, *CP* crude protein, *EE* ether extract, *WSC* water-soluble carbohydrate

[a]The numbers are $\log_{10}$ and the unit is colony-forming unit/g

the CP content of the ensiled material by ~5.7%, which is within normal levels of expectation.

Crude protein hydrolyses may occurs inside silage clamp during fermentation period which might be due to either the activity of proteolysis plant enzymes and/or plant endogenous micro-organisms, resulting a reduction in CP and increase in NH_3-N concentration of produced silage [1, 6]. Water-soluble carbohydrate values were found to be significantly lower in the ensiled material but treating the fresh, MG with EM Additive had no effect on the use of WSC as a substrate for microbial activity during the ensiling process. Wilkinson and Davies [6] reported that during the fermentation process at ensiling most of forage WSC convert to fermentation acids mainly lactic and butyric acids. This could be the reason for reducing the WSC concentration at ensiling samples from 72.9 g/kg DM inside fresh samples to 12.1 g/kg DM inside ensiled samples of our study. Using the mix of two strains of LAB and *S. cerevisiae* had no effect on all studied proximate analysis in the present study (Table 2). Same results were found by several researchers [9, 21–23]. While other studies found that LAB reduced ($P < 0.05$) WSC concentration, which could be that most of WSC were fermented to acids compared to untreated forage silages [24, 25].

3.2 Fermentation Characteristics

Meadow grass samples were found to ensiled well as indicated by a low mean pH of ~4.14 and NH_3-N ~1.2 g/kg TN concentration, which refer to a good ensiled technique. However, using EM treatment as a silage additive had no effect on the final silage pH. Using different LAB strains and *S. cerevisiae* as silage inoculate showed a trend ($P = 0.09$) to reduce silage pH with no ($P > 0.05$) effect on NH_3-N. Shah et al. [23] found that using inoculate (LAB) as silage additives significantly reduced silage pH, acetic acid and butyric acid while silage inoculate had no effect on DM content of produced silage. Amanullah et al. [9] reported that using homo-fermentative LAB rapidly dropped silage pH and reduced the concentration of NH_3-N. Duniere et al. [8] noticed the highest concentration of lactic acid and lowest pH when they used LAB as inoculated bacteria after 7, 14, 28 and 56 days of ensiling whole crop corn silage.

Treating the fresh, chopped MG sample with EM additive resulted in a reduction (~13%) of lactic acid in the final silage. The results obtained from this research showed that using EM did not show any negative effect on fermentation products or nutrient content of produced silage. A similar result was observed by Duniere et al. [8] when they used three different strains of Saccharomyces as silage inoculate to treat corn silage after 90 of the ensiling periods. Pedroso et al. [26] reported that *Saccharomyces* spp. can convert WSC into CO_2 and ethanol under anaerobic condition. These products are undesirable product inside silage and could be toxic to ruminant animals. This could be the reason of increasing the level of ethanol when grass treated with a mix of yeast and inoculation bacteria in the current study

Table 2 Effect of using three microbial strains as a silage additive on the silage fermentation and aerobic stability of meadow grass

Parameter	Control	EM treated	SED	LSD	*P*-value
pH	4.15	4.133	0.091	0.203	0.858
Ammonia g/kg	1.12	1.22	0.0581	0.1295	0.133
Ammonia %TN	1.663	1.771	0.1501	0.334	0.486
Lactic acid g/kg DM	34.55	30.11	0.812	1.81	<0.001
Total VFA [t VFA] mg/kg	9287	12,427	627	1397.1	<0.001
Acetic acid mg/kg	8775	12,090	773.5	1723.5	0.002
Propionic acid mg/kg	48.7	92.2	17.45	38.87	0.032
Iso-Butyric acid mg/kg	46.7	34.8	15.49	34.51	0.463
Butyric acid mg/kg	52.2	36.3	16.34	36.4	0.355
Iso-Valeric acid mg/kg	0.6	0.4	0.2	0.45	0.213
Valeric acid mg/kg	197	42	148.9	331.9	0.324
Hexanoic acid mg/kg	57	44	25.2	56.1	0.635
Heptanoic acid mg/kg	54	44	25.3	56.4	0.696
Total alcohol mg/kg	2547	8350	324.5	723	<0.001
Ethanol mg/kg	1661	4041	196.7	438.3	<0.001
Propan-1-ol mg/kg	26.7	105.2	5.2	11.58	<0.001
Propan-1,2-diol mg/kg	860	4204	255.4	569	<0.001
Ethanol %total alcohol	65.5	48.5	2.67	5.94	<0.001
Propan-1-ol %total alcohol	1.1	1.3	0.11	0.24	0.101
Propan-1,2-diol %total alcohol	33.5	50.2	2.68	5.97	<0.001
Aerobic stability[a]	6.83	13.79	1.207	2.782	<0.001

EM experimental treatment, *TN* total nitrogen, *DM* dray matter, *VFA* volatile fatty acid
[a]Number of days to reach 3 °C above ambient

(Table 2). In contrast to our results, Duniere et al. [8] noted that the level of ethanol did not increase by using a mix of *Saccharomyces* spp. The concentration of acetic acid, propionic acid, ethanol and total and individual alcohols were found to be increased by silage inoculate that used on our study (Table 2). The formation of acetic acid and 1,2-propanediol from lactic acid might be due to the activity of *L. buchneri* and would lead to increase the formation of CO_2 and this would lead to extra DM losses [10]. Thus, the DM losses in the current study would be due to hetro-fermentative that appear as a result of the activity of *L. buchneri*. Shah et al. [23] and Zhang et al. [25] also observed that using inoculated bacteria enhanced ethanol concentration in produced silage either after 30 or after 100 days.

Mould and yeast counts were found to be affected by the ensiling process, ensiled MG had higher mould and lower yeast counts compared to fresh forages. However, the use of the EM additive had no effect on the mean of mould or yeast (Table 1).

Lactobacilli numbers were found to be significantly lower (~48.9%) in the MG sample treated with the EM additive. Yeasts and mould contamination either for fresh or ensiled forage were generally in low level (>3.2 log10 CFU/g) when compared to yeasts and mould contaminations with other studies [10, 19, 27]. The effect of fermentation process on yeast counting could be due to the reduction of silage pH which has an inhibitory effect on yeast counting, hence decrease their number in ensiled MG. Driehuis et al. [27] showed that the presence of *L. buchneri* inside silage silo can convert part of fermented lactic acid to acetic acid and 1,2-propanediol which has an inhibitory effect on yeasts and mould during primary fermentation under anaerobic conditions. Conversely adding the EM treatment to the fresh MG sample resulted in an elevated total volatile fatty acid concentration by ~34%. Elevated tVFA related to a ~38% increase in acetic acid and ~89% increase in propionic acid as shown in Table 2. No other VFA were affected by the EM treatment and the relative proportion of the individual VFA. The total alcohol concentration of the ensiled material was found to increase (~228%) as a result of EM treatment. Elevated alcohol concentration related to increases in ethanol (~143%), propane-1-ol (~294%) and propane-1,2-diol (~389%). The relative proportions of the different alcohols were also found to be affected by EM treatment. The addition of EM additive significantly reduced the relative proportion of ethanol whilst increased the relative proportion of propane-1,2,-diol as shown in Table 2.

4 Aerobic Stability

The addition of EM treatment at ensiling resulted in a more aerobically stable silage. Aerobic stability result from the current study showed that untreated silage samples needed (6.83 days) to increase the silage temperature 3 °C above ambient (Table 2 and Fig. 1), while treating grass silage with EM was increased the time required approximately 100% to reach the same temperature above ambient, which give an indication the EM had enhanced aerobic stability of produced silage. Li et al. [28] reported that aerobic spoilage occurred within 7 days after air exposure. Silage spoilage occurs when the silage exposed to oxygen and the activity of undesirable aerobic microorganisms is started, hence the pH and the silage temperature will increase [1]. Spoiled silage is undesirable for feeding animals, and this could be attributed to that, spoiled silage has poorer quality, lower digestibility and high risk of diseases and hence has a negative effect on animal performance. Several studied [20, 23, 29, 30] reported that silage aerobic stability was enhanced when the different stain of LAB alone or mixed were used as silage additive. Opening the silage silo and exposure to air can enhanced yeast growth, reduce the level of lactic acid and increasing silage pH, and tolerate aerobic microbes to grow such as mould causing silage deterioration [8]. A similar result was noticed by Duniere et al. [8] who reported that using a mix of different LAB species or homofermentative LAB were reduced the susceptibility of grass silage to aerobic deterioration.

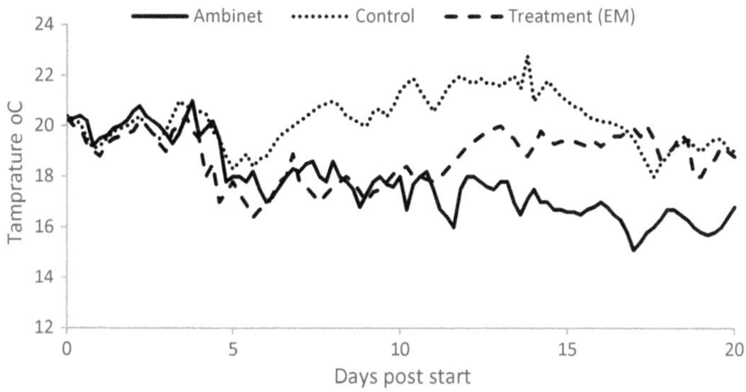

Fig. 1 Effect of using three microbial strains as a silage additive on the silage aerobic stability temperature of meadow grass, EM: experimental treatment

5 Conclusion

Meadow grass samples ensiled well in the mini-silos resulting in a final pH of ~4.1 with very low NH3-N%TN values. Using a mix of *S. cerevisiae*, *L. plantarum* and *L. casei* treatment elevated LAB populations [log10 CFU/g] but resulted in a reduction lactic acid concentration. Elevated VFA and alcohol concentrations were observed as a result of EM treatment. EM treatment resulted in a more aerobically stable PRG silage possibly as a result of the elevated VFA and alcohol concentrations [ethanol & propane-1,2-diol].

References

1. Merry, R.J., Davies, D.R.: Propionibacteria and their role in the biological control of aerobic spoilage in silage. Le Lait **79**(1), 149–164 (1999)
2. Jones, R.: Understanding the Processes of Protein Degradation in Forage Crops Provides Opportunities for Improved Silage Quality and Enhanced Animal Production. Institute of Grassland and Environmental Research, UK (Courtesy of Alltech Inc.) (2007)
3. Pitt, E.R. and Shaver, R.D.: Proceedings of Dairy Feeding Systems Symposium, Harrisburg, USA (1990)
4. Hart, K.J.: Evaluation of Ensiled Whole-Crop Legumes for Ruminants. Open University (2005)
5. Taha, V.: Effect of Supplemental Tannin on Silage Quality and Animal Performance. Harper Adams University (2015)
6. Wilkinson, J., Davies, D.: The aerobic stability of silage: key findings and recent developments. Grass Forage Sci. **68**(1), 1–19 (2013)
7. Salawu, M.B., Warren, E.H., Adesogan, A.T.: Fermentation characteristics, aerobic stability and ruminal degradation of ensiled pea/wheat bi-crop forages treated with two microbial inoculants, formic acid or quebracho tannins. J. Sci. Food Agric. **81**(13), 1263–1268 (2001)
8. Duniere, L., Jin, L., Smiley, B., Qi, M., Rutherford, W., Wang, Y., McAllister, T.: Impact of adding *Saccharomyces* strains on fermentation, aerobic stability, nutritive value, and select lactobacilli populations in corn silage. J. Anim. Sci. **93**(5), 2322–2335 (2015)

9. Amanullah, S., Kim, D., Lee, H., Joo, Y., Kim, S., Kim, S.: Effects of microbial additives on chemical composition and fermentation characteristics of barley silage. Asian-Australas. J. Anim. Sci. **27**(4), 511 (2014)
10. Kim, D.H., Amanullah, S.M., Lee, H.J., Joo, Y.H., Han, O.K., Adesogan, A.T., Kim, S.C.: Effects of hybrid and bacterial inoculation on fermentation quality and fatty acid profile of barley silage. Anim. Sci. J. **89**(1), 140–148 (2018)
11. Davies, D., Merry, R., Williams, A., Bakewell, E.L., Leemans, D., Tweed, J.: Proteolysis during ensilage of forages varying in soluble sugar content. J. Dairy Sci. **81**(2), 444–453 (1998)
12. Lettat, A., Nozière, P., Silberberg, M., Morgavi, D.P., Berger, C., Martin, C.: Rumen microbial and fermentation characteristics are affected differently by bacterial probiotic supplementation during induced lactic and subacute acidosis in sheep. BMC Microbiol. **12**(1), 142 (2012)
13. MAFF: The analysis of agriculture materials. In: Edition, r. (ed.) Food, Reference Book: Manual of Veterinary Parasitological Laboratory Techniques. Ministry of Agriculture and Fisheries, HMSO, London (1986)
14. Horwitz, W.: Official Methods of Analysis of AOAC International, 17th edn. Association of Analytical Chemists International, Gaithersburg, MD (2000)
15. Borreani, G., Tabacco, E.: Improving corn silage quality in the top layer of farm bunker silos through the use of a next-generation barrier film with high impermeability to oxygen. J. Dairy Sci. **97**(4), 2415–2426 (2014)
16. Canale, A., Valente, M.E., Ciotti, A.: Determination of volatile carboxylic acids (C1–C5i) and lactic acid in aqueous acid extracts of silage by high performance liquid chromatography. J. Sci. Food Agric. **35**(11), 1178–1182 (1984)
17. Cruwys, J., Dinsdale, R., Hawkes, F., Hawkes, D.: Development of a static headspace gas chromatographic procedure for the routine analysis of volatile fatty acids in wastewaters. J. Chromatogr. A **945**(1–2), 195–209 (2002)
18. Mayne, C., Gordon, F.: Effect of harvesting system on nutrient losses during silage making. 2. In-silo losses. Grass Forage Sci. **41**(4), 341–351 (1986)
19. Wambacq, E., Latré, J.P., Haesaert, G.: The effect of Lactobacillus buchneri inoculation on the aerobic stability and fermentation characteristics of alfalfa-ryegrass, red clover and maize silage. Agric. Food Sci. **22**(1), 127–136 (2013)
20. Reich, L.J., Kung Jr., L.: Effects of combining *Lactobacillus buchneri* 40788 with various lactic acid bacteria on the fermentation and aerobic stability of corn silage. Anim. Feed Sci. Technol. **159**(3–4), 105–109 (2010)
21. Zahiroddini, H., Baah, J., McAllister, T.: Effects of microbial inoculants on the fermentation, nutrient retention, and aerobic stability of barley silage. Asian Australas. J. Anim Sci. **19**(10), 1429–1436 (2006)
22. Baah, J., Addah, W., Okine, E., McAllister, T.: Effects of homolactic bacterial inoculant alone or combined with an anionic surfactant on fermentation, aerobic stability and in situ ruminal degradability of barley silage. Asian Australas. J. Anim. Sci. **24**(3), 369–378 (2011)
23. Shah, A., Xianjun, Y., Zhihao, D., Siran, W., Tao, S.: Effects of lactic acid bacteria on ensiling characteristics, chemical composition and aerobic stability of king grass. J. Anim. Plant Sci. **3**, 747–755 (2017)
24. Owens, V., Albrecht, K., Muck, R.: Protein degradation and fermentation characteristics of unwilted red clover and alfalfa silage harvested at various times during the day. Grass Forage Sci. **57**(4), 329–341 (2002)
25. Zhang, L., Yu, C., Shimojo, M., Shao, T.: Effect of different rates of ethanol additive on fermentation quality of Napiergrass (*Pennisetum purpureum*). Asian Australas. J. Anim. Sci. **24**(5), 636–642 (2011)
26. Pedroso, A.d.F., Nussio, L.G., Paziani, S.d.F., Loures, D.R.S., Igarasi, M.S., Coelho, R.M., Packer, I.H., Horii, J., Gomes, L.H.: Fermentation and epiphytic microflora dynamics in sugar cane silage. Sci. Agricola **62**(5), 427–432 (2005)
27. Driehuis, F., Elferink, S.O., Van Wikselaar, P.: Fermentation characteristics and aerobic stability of grass silage inoculated with *Lactobacillus buchneri*, with or without homofermentative lactic acid bacteria. Grass Forage Sci. **56**(4), 330–343 (2001)

28. Li, Y., Wang, F., Nishino, N.: Lactic acid bacteria in total mixed ration silage containing soybean curd residue: their isolation, identification and ability to inhibit aerobic deterioration. Asian Australas. J. Anim. Sci. **29**(4), 516 (2016)
29. Holzer, M., Mayrhuber, E., Danner, H., Braun, R.: The role of *Lactobacillus buchneri* in forage preservation. Trends Biotechnol. **21**(6), 282–287 (2003)
30. Schmidt, R., Hu, W., Mills, J., Kung Jr., L.: The development of lactic acid bacteria and *Lactobacillus buchneri* and their effects on the fermentation of alfalfa silage. J. Dairy Sci. **92**(10), 5005–5010 (2009)

Extracting Cellulose Fibers from Rice Husks to Prepare a pH Sensitive Hydrogel with Sodium Alginate

Alarqam Zyaad Tareq, Mohammed Salim Hussien, Asaad Mohammed Mustafa and Ahmed Raoof Mahmood

Abstract In this paper, pH-sensitive hydrogels were prepared and developed from cellulose fibers extracted from rice husks with sodium alginate by using a double cross-linker. Cellulose fibers extraction process was carried out by alkaline and bleaching treatment. Cellulose fibers were used in variable quantities to investigate their effects on swelling degree. The resulted hydrogels showed a very strong swelling (1785–2718%) in phosphate buffer solution (pH 7), good swelling (776–1195%) in pH 10, and less swelling (81–124%) in pH 4. All of these swelling degrees and more properties of the resulted cellulose fibers and hydrogels were tested and evaluated via some techniques like Infrared spectroscopy (ATR), thermal analysis (TG, DTG, DSC, and DTA), and Scanning electron microscopy (SEM). These results show that cellulose/alginate hydrogels could provide many possible applications in the biomedical, and purification of the wastewater by adsorbing pollutants.

Keywords Cellulose fibers · pH-sensitive · Hydrogel · Rice husks · Sodium alginate

A. Z. Tareq (✉) · M. S. Hussien · A. M. Mustafa
University of Zakho, Dahuk, Iraq
e-mail: Alarqam.tareq@uoz.edu.krd

M. S. Hussien
e-mail: mohammed.huseein@uoz.edu.krd

A. M. Mustafa
e-mail: Asaad.mustafa@uoz.edu.krd

A. R. Mahmood
Imam Ja'far Al-Sadiq University, Kirkuk, Iraq
e-mail: ahmedbayatil89@gmail.com

Y. T. Mustafa et al. (eds.), *Recent Researches in Earth and Environmental Sciences*,
Springer Proceedings in Earth and Environmental Sciences,
https://doi.org/10.1007/978-3-030-18641-8_8

1 Introduction

Rice husks are defined as a one of the main agriculture residues. They are mostly generated during the rice milling process as a secondary product. These agricultural residues represent 23% of the 649.7 million ton of rice that are produced per year in the world [1]. It is very familiar that this natural polymer, Cellulose, is widely abundant in this universe. Therefore, it is used a lot in industries to leave behind a big amount of residues difficult to be disposed or reused. For instance, in the milling process of cereal crop such as rice, the generated cellulosic waste is discarded in soil [2].

Cellulose chemically structured of a long chain polymer that consists of D-glucose as a repeating units, called pyranose. These pyranose units have acetyl linkages to join the carbon 1 of one pyranose ring with the carbon 4 of the following ring by one oxygen atom, called β-1-4 linkages [3]. The physical properties of cellulose are affected by intra and inter molecular hydrogen bonds which are formed by hydroxyl groups of β-1-4-glucopyranose [4]. Sodium alginate is structurally linear-chains polysaccharide. It consists of alternating blocks of 1–4 linked α-L-guluronic acid and β-D-mannuronic acid residues [5]. Sodium alginate has been listed as a safe substance by the Food and Drug Administration (FDA) [6].

Hydrogels are categorized as hydrophilic polymers with three-dimensional network structure. They have the ability of absorbing and retaining a big amount of water. Because of this distinct feature besides other features like excellent hydrophobicity, permeability, compatibility, and low coefficient of friction, they have been mainly used in the pharmaceutical purposes, drag delivery system, agriculture, and food industry [7–10].

Going further in preparing hydrogel networks, it was necessary to use one of the crosslinking processes: chemical or physical. Chemical crosslinking network is characterized by having permanent junctions; while the physical is characterized by having junctions forces that formed from either polymer chain entanglements or physical interactions such as Ionics interactions, and hydrogen bonds [11]. The combination of a variety of biopolymers is an inexpensive, extremely significant, and useful way used to form new structural materials. Thus, cellulose/biopolymer composite hydrogels specifically have recently very remarkable attention; Wang et al. prepared cellulose/collagen hydrogel beads for Cu (II) adsorption [12] and Peng et al. reported the immobilization of laccase on cellulose/chitosan beads [13]. All of these are indicating that cellulose-based hydrogels blended with various biopolymers can contribute as novel materials for many applications.

Thus, the aim of this work was to extract and investigate cellulose fibers from rice husks to prepare cellulose/sodium alginate hydrogels in various ratios by using double crosslinking including glutaraldehyde and calcium chloride aqueous solution. The cellulose/alginate hydrogels were investigated by Scanning Electron Microscope (SEM), Infrared spectroscopy (ATR), swelling behavior, and thermal stability (TG, DTG, DSC, and DTA).

2 Materials and Methods

The rice husks, used as a raw material, were obtained from Akre, Kurdistan region, Iraq. Sodium alginate (medium viscosity), sodium chloride, acetic acid, calcium chloride, sodium chlorite, and sodium hydroxide (purity 99%) were purchased from BDH Chemical company, UK. Glutaraldehyde was obtained from Merck KGaA Chemical company, USA. All other chemicals were used without any further purification.

2.1 Extracting Cellulose Fibers from Rice Husks

The rice husks were washed by D.W. and dried in convection oven at 70 °C for 24 h. Blander was used to grind the sample until husks fibers were formed in a specific size (Fig. 1a). The extraction of cellulose fiber was performed by using alkaline treatment to be followed by bleaching treatment.

2.2 Alkaline Treatment

The aim of the alkaline process was to remove lignin and hemicellulose from rice husks to get pure cellulose [14, 15]. The rice husks were treated with an alkali solution (4 wt% NaOH). Then, the solution was moved to around bottom flask. The whole process was carried out at reflux temperature for 2 h. A filtration and washing with D.W. were carried out to the mixture several times. This treatment was repeated three times. The product, shown in Fig. 1b, was dried before using it for bleaching treatment.

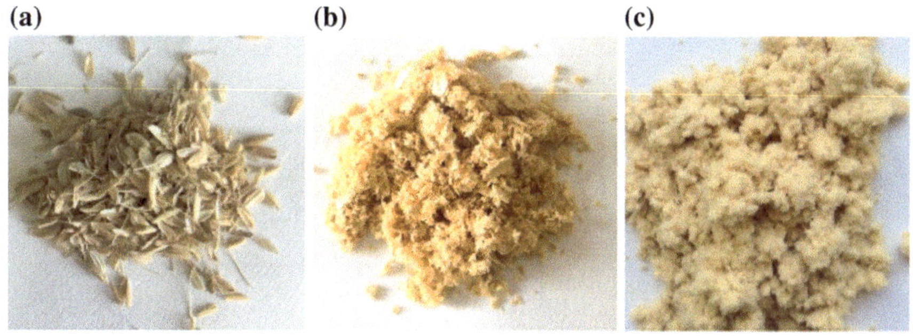

Fig. 1 Images of rice husk: **a** untreated rice husk, **b** alkali-treated rice husk, and **c** bleached rice husk

2.3 Bleaching Treatment

After the previous treatment, a process of bleaching was conducted by adding a solution of acetic acid, aqueous solution of chlorine (1.7 wt%) with D.W. at reflux at 110–130 °C (using oil bath) for four hours. The resulted solution was left to cool, then filtered and washed with excess D.W. The bleaching treatment was executed 4 times and the final product is shown in Fig. 1c.

2.4 Preparation of Cellulose/Sodium Alginate Hydrogels

Cellulose fibers were obtained from rice husks and sodium alginate were used in the synthesis of hydrogels. Some of sodium alginate was dissolved in D.W. at 30 ± 1 °C with stirring (150 rpm) until a 10% of clear solution was obtained. Different amounts of cellulose fibers (according to Table 1) were dispersed in D.W. with stirring at 35 °C for 1 h, then, they were gently added to the previously prepared sodium alginate solution. An amount (20 mg/ml D.W.) of calcium chloride was used with stirring (200 rpm) for 30 min at 30 °C. The following step was the immerse of the mixture in $CaCl_2$ solution for 24 h at 35 °C, and then the result was washed with warm D.W. and immersed in 0.5% glutaraldehyde solution for 2 h at 35 °C. The formed hydrogels, shown in Fig. 2, were washed by D.W. several times and dried at 40 °C, as in Fig. 2.

Table 1 Combination of cellulose/alginate hydrogel

No. of sample	Sodium alginate (g)	Cellulose fibers (g)
1	0.5	0.005 (1%)
2	0.5	0.01 (2%)
3	0.5	0.02 (4%)
4	0.5	0.03 (6%)
5	0.5	0.04 (8%)
6	0.5	0.05 (10%)

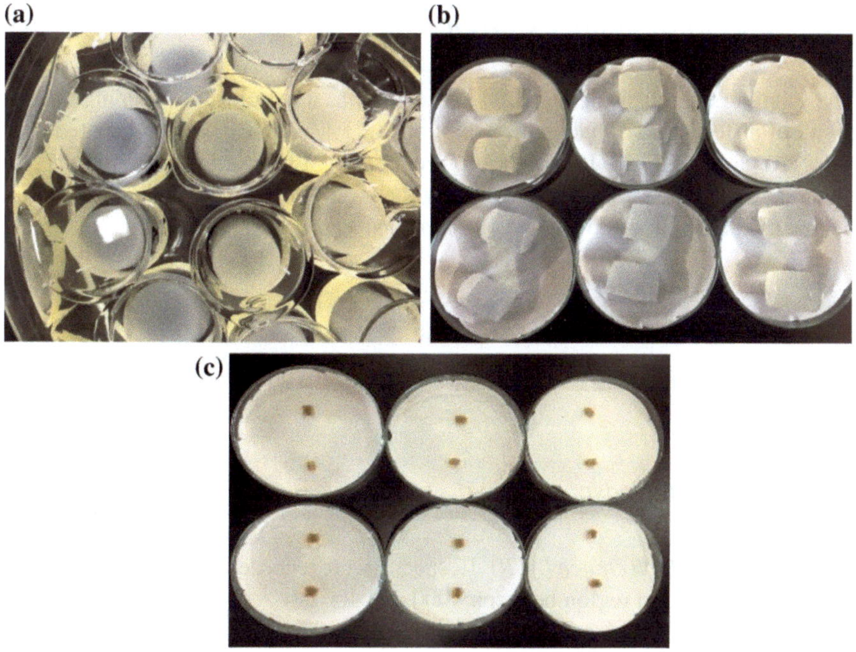

Fig. 2 Cellulose/alginate hydrogel **a**, **b** after swelling **c** before swelling

2.5 *Scanning Electronic Microscopy (SEM)*

The morphology of the cellulose fibers and hydrogel prepared was evaluated by using Scanning Electron Microscopy (SEM) ultra-high resolution (TESCAN MAIA3, Czech Republic, France) with accelerated electrons of 5 kV of energy.

2.6 *Thermal Analysis*

Thermogravimetric analysis (TG), Differential Scanning Calorimetry (DSC), Differential Thermogravimetric analysis (DTG), and Differential Thermal Analysis (DTA) were performed to study the decomposition properties of the prepared hydrogel. The thermal stability of hydrogel was determined by using Hitachi STA 7300, Japan. Five mg of sample was analyzed in nitrogen atmosphere with 50 ml/min of gas flow rate, 10 °C/min of heating rate, and at a range of 50–800 °C of temperature.

2.7 *Study of Swelling Behavior*

The swelling degree of the prepared hydrogels was determined by keeping them in 50 ml of different phosphate buffer solutions (pH 4, 7, and 10) and incubated in an autoclave at 37 °C. The increase of hydrogel weight ($W_t - W_0$) at different time intervals was compared to the initial weight (W_0) in order to calculate the swelling degree (S_w) as in the following equation below:

$$(S_w\,\%) = \frac{W_t - W_0}{W_0} * 100 \qquad (1)$$

where W_0 and W_t are the initial and final weight of hydrogel respectively.

2.8 *FT-IR Spectroscopy*

Attenuated Total Reflectance (ATR) IR spectra were recorded by using Bruker Alpha Platinum in a spectra region between 4000 and 400 cm^{-1} at a resolution of 4 cm^{-1} to assess the functional group of cellulose fibers, sodium alginate, and hydrogel prepared.

3 Results and Discussion

3.1 *IR Spectroscopy Analysis*

The IR technique was used to assess the main functional groups that are present in sodium alginate, cellulose fibers, and the prepared hydrogel. IR spectra of sodium alginate are realized in Fig. 3a which shows a strong peak at 1405 and 1592 cm^{-1} due to asymmetric and symmetric stretching of carboxylate salt groups. Another broad peak was seen at 3281 cm^{-1}, which is attributed to the –OH group. The asymmetric stretching peaks at 1081–1012 cm^{-1} corresponding to C–O–C group [16]. IR spectra of the cellulose fibers are shown in Fig. 3b. The broad band in rang of 3100–3400 cm^{-1} was attributed to the –OH stretching vibrations of cellulose [17], and a peak of 2256 cm^{-1} referred to C–H stretching vibrations. The peak of 1646 cm^{-1} could explain the bending of the –OH groups of cellulose. At the meanwhile, other absorption peaks were observed from 1800 to 500 cm^{-1} to interpret the deformation, wagging and twisting modes of the an hydro-glucopyranose units [18]. Moreover, the disappearance or changes in the intensity of certain IR peaks of the crystalline domain of cellulose are related to the change of the cellulose crystal

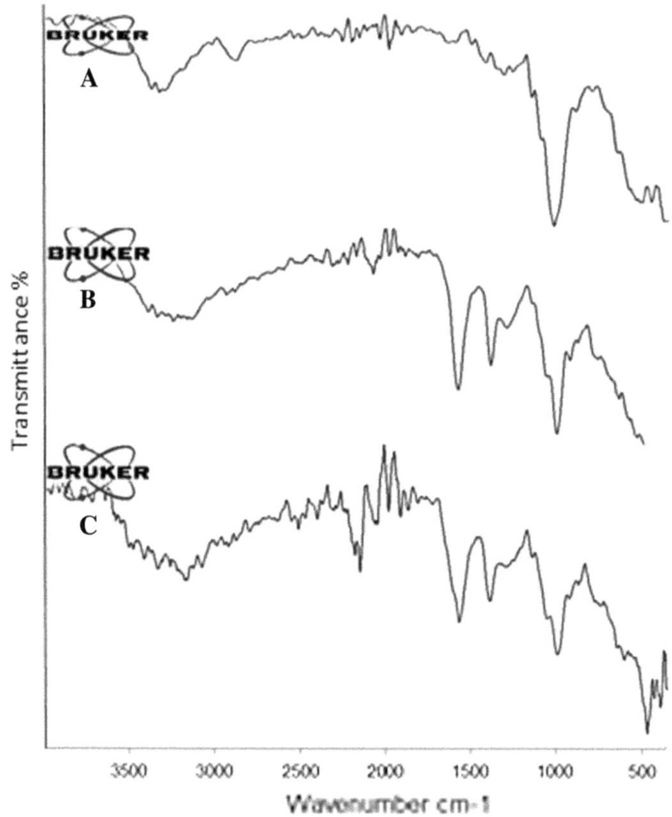

Fig. 3 IR spectra of **a** cellulose fibers, **b** sodium alginate, **c** cellulose/alginate hydrogel

structure [19]. The preparation of cellulose-alginate hydrogel was followed by a slight decrease in the intensity of the peak at 3222–3666 cm^{-1}, which was due to the –OH vibrations, to illustrate the cross-linking between cellulose and alginate.

3.2 *Study of Swelling Behavior*

Figure 4a–c show the degree of swelling of cellulose-alginate hydrogels in different phosphate buffer solutions (pH 4, 7, and10) respectively. Then, the effect of pH solutions of the % swelling of the hydrogels was observed.

The obtained results were pointing out that the swelling behavior increased with the pH increase of the swelling medium for the hydrogels prepared. This increase

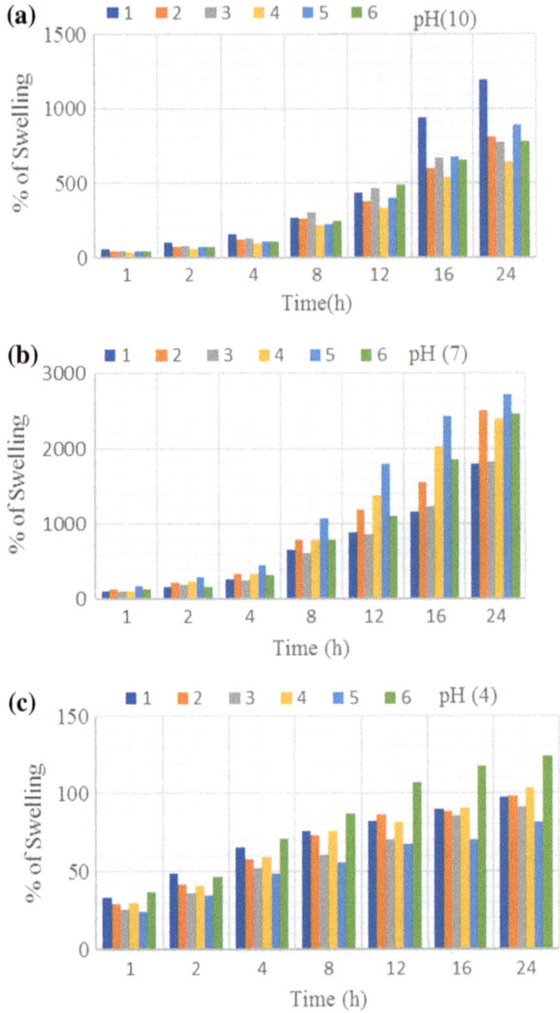

Fig. 4 Swelling behavior of cellulose/alginate hydrogel in different phosphate buffer solution pH (4, 7, and 10)

in swelling % could interpret the fact that with rising the pH of swelling medium, the ratio –COO–/–COOH on cellulose fibers also was increasing the ionization of carboxylic groups. This fact led to a greatest repulsion among the –COO– earing cellulose chains. The results could also be explaining the fact that the more increase in the swelling medium pH, the more increase in the extent of ionization of carboxylate groups of alginate. This alginate provides a large number of carboxylate ions through alginate molecules. These anionic charged centers alienated each other to produce a rapid relaxation in the hydrogel network. This led obviously to a rise in the swelling degree.

3.3 Morphology Analysis

The images of rice husks fibers after each phase of treatment is shown and illustrated in Fig. 1a–c. It is appeared that the color of the untreated rice husks is yellow and turns to be light yellow after alkali treatment. Then, the bleached material is appeared differently in excessive fibrous structure with very light-yellow color. The variation in colors is attributed to the remove of non-cellulosic materials like lignin, hemicellulose, pectin and wax within the chemical treatment of the rice husks. The resulted light-yellow color with fibrous structure of the final product can refer obviously to the pure cellulosic fibers.

At the meanwhile, Fig. 5a–c shows the scanning electron microscopy (SEM) images and the cross-section of the cellulose fibers resulting from bleaching treatment of rice husks. It was clear that the cellulose fibers had a long fibrous structure with curled and soft-flat shape. The surface was rough with pits. The average width was 4.5 μm, which was reported by Kumar et al. [20]. The curled shape increased the surface area and made the fibers more reactive, while the flat shape of these fibers increased their specific area.

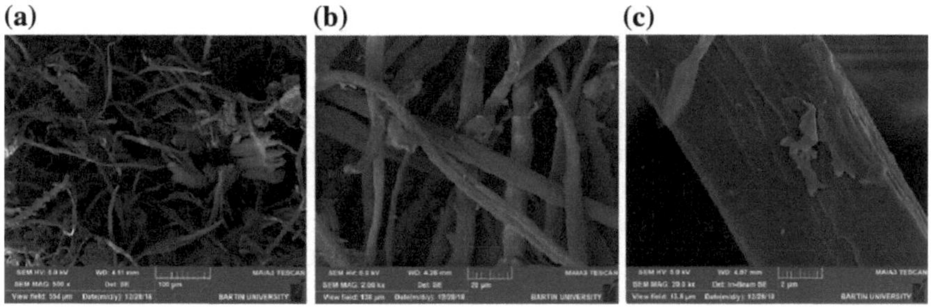

Fig. 5 Scanning electron microscopy of cellulose fibers

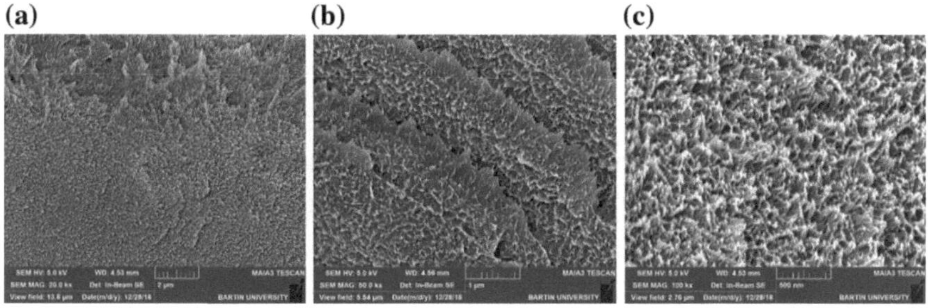

Fig. 6 Scanning electron microscopy of cellulose/alginate hydrogel

Figure 6a–c, however, shows the SEM images of the surface and cross-section of cellulose-alginate hydrogel. The observation of the SEM image (a) indicated that the cellulose fibers were non-homogenously distributed and well blended with alginate in the cellulose-alginate hydrogel. Furthermore, they indicated for a good integration between the alginate and cellulose fibers. The cross-section (b) shows the roughness with layer-shaped folds similar to a leaf. The other cross-section (c), at higher magnification, shows non-smooth or rough surface with deep holes enable the solutions to penetrate into the hydrogel structure, and no aggregated cellulose fibers. The rough surface of the hydrogel was attributed to the non-homogenous dispersion of the cellulose fibers in the polymer matrix.

3.4 *Thermal Analysis*

Thermal stability of the cellulose-alginate hydrogel was evaluated by TG, DTG, DSC, and DTA and the results are shown in Table 2 and Fig. 7.

TG curve shows the weight of decreasing pattern of the hydrogel sample which is attributed to the thermal decomposing. DTG curve shows the rate of the thermal decomposing. DSC and DTA curves show the enthalpy of fusion (ΔH) and maximum melting temperature.

Thermal decomposing of the hydrogel sample was observed generally in four steps. At 187.9 °C the initial weight loss was caused by evaporating moisture and started breaking the outer structure. The following step of thermal decomposing was at 461.3 °C, which means that polymer chains lost their compactness and became more brittle. At the third and four steps, the decomposition of the cellulosic and alginate materials was observed at a range of 607–693 °C. The DSC and DTA curves illustrate the exothermic analysis. The melting temperature™ of the hydrogel was elevated to be equal to the maximum temperature of the decomposing (Tmax). The latter is an indication to an increase in melting temperature as a result to the formation of new functional groups along the polymer chains creating intermolecular forces that shift the hydrogel Tm to that of Tmax.

Table 2 Thermal analysis details of cellulose/alginate hydrogel

DTG		DSC and DTA		TG	
Tmax °C	Rate of decomp. (μg/min)	Tm °C	Enthalpy of fusion ΔHm (mJ/mg)	Initiate temp. °C	Terminate temp. °C
500	397.5	500	Exotherm (−540)	187.7	692.8

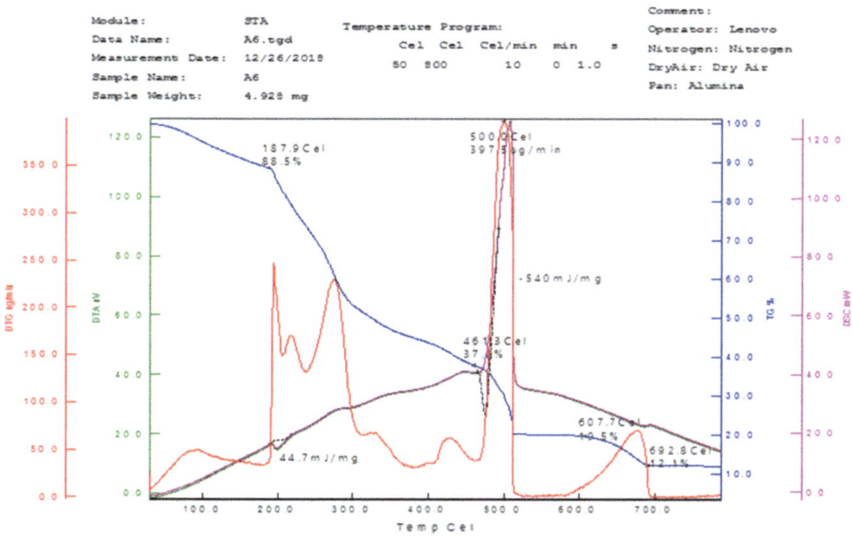

Fig. 7 Thermal analysis diagrams of cellulose/alginate hydrogel

4 Conclusion

Cellulose fibers were successfully isolated from rice husks by alkaline process and bleaching treatment. pH sensitive hydrogels were obtained from cellulose fibers and sodium alginate by using double cross-linking (glutaraldehyde and calcium chloride). The cellulose fibers show high purity degree by using IR techniques with SEM to study the fibers structure. Additionally, the hydrogels prepared show a homogenous pores investigated by SEM, a strong sensitivity in pH medium (swelling behavior) with better thermal properties investigated by TG, DTG, DSC, and DTA.

5 Future Work

To purify wastewater from pollutants, the cellulose/alginate hydrogel can be used affectively depending on its pH-sensitive in the solutions.

References

1. Chandrasekhar, S., Pramada, P.N., Majeed, J.: Effect of calcination temperature and heating rate on the optical properties and reactivity of rice husk ash. J. Mater. Sci. **41**(23), 7926–7933 (2006). https://doi.org/10.1007/s10853-006-0859-0
2. Chandrasekhar, S., Satyanarayana, K., Pramada, P., Raghavan, P., Gupta, T.: Review processing, properties and applications of reactive silica from rice husk—an overview. J. Mater. Sci. **38**(15), 3159–3168 (2003)

3. Kim, J., Yun, S., Ounaies, Z.: Discovery of cellulose as a smart material. Macromolecules **39**(12), 4202–4206 (2006)
4. Zain, N.F., Yusop, S.M., Ahmad, I.: Preparation and characterization of cellulose and nanocellulose from pomelo (*Citrus grandis*) albedo. J. Nutr. Food Sci. **5**(1), 334 (2014)
5. Tareq, A.Z.: Preparation and characterization of PVA-NaAlg interpenetrating polymer network (IPN) hydrogel for controlled delivery of carbidopa. Am. Chem. Sci. J. **17**(1), 1–10 (2016). https://doi.org/10.9734/acsj/2016/28990
6. Tareq, A.Z., Hussein, M.S.: Sodium alginate-gelatin cross-linked microspheres for releasing diltiazem HCl. Sci. J. Univ. Zakho **4**(2), 226–235 (2016)
7. Tareq, A.Z., Hussein, M.S., Mustafa, A.M.: Synthesis and characterization of PVA-gelatin hydrogel membranes for controlled delivery of captopril. Int. Res. J. Pure Appl. Chem. **12**(4), 1–10 (2016). https://doi.org/10.9734/IRJPAC/2016/28989
8. Prabaharan, M., Mano, J.F.: Stimuli-responsive hydrogels based on polysaccharides incorporated with thermo-responsive polymers as novel biomaterials. Macromol. Biosci. **6**(12), 991–1008 (2006)
9. Cai, J., Zhang, L.: Unique gelation behavior of cellulose in NaOH/urea aqueous solution. Biomacromol **7**(1), 183–189 (2006)
10. Söderqvist Lindblad, M., Albertsson, A.-C., Ranucci, E., Laus, M., Giani, E.: Biodegradable polymers from renewable sources: rheological characterization of hemicellulose-based hydrogels. Biomacromolecules **6**(2), 684–690 (2005). https://doi.org/10.1021/bm049515z
11. Rodríguez, R., Alvarez-Lorenzo, C., Concheiro, A.: Cationic cellulose hydrogels: kinetics of the cross-linking process and characterization as pH-/ion-sensitive drug delivery systems. J. Control. Release **86**(2), 253–265 (2003). https://doi.org/10.1016/S0168-3659(02)00410-8
12. Wang, J., Wei, L., Ma, Y., Li, K., Li, M., Yu, Y., Wang, L., Qiu, H.: Collagen/cellulose hydrogel beads reconstituted from ionic liquid solution for Cu (II) adsorption. Carbohydr. Polym. **98**(1), 736–743 (2013)
13. Zheng, F., Cui, B.-K., Wu, X.-J., Meng, G., Liu, H.-X., Si, J.: Immobilization of laccase onto chitosan beads to enhance its capability to degrade synthetic dyes. Int. Biodeterior. Biodegradation **110**, 69–78 (2016)
14. Abe, K., Yano, H.: Comparison of the characteristics of cellulose microfibril aggregates of wood, rice straw and potato tuber. Cellulose **16**(6), 1017–1023 (2009). https://doi.org/10.1007/s10570-009-9334-9
15. Chen, W., Yu, H., Liu, Y., Chen, P., Zhang, M., Hai, Y.: Individualization of cellulose nanofibers from wood using high-intensity ultrasonication combined with chemical pretreatments. Carbohydr. Polym. **83**(4), 1804–1811 (2011). https://doi.org/10.1016/j.carbpol.2010.10.040
16. Lawrie, G., Keen, I., Drew, B., Chandler-Temple, A., Rintoul, L., Fredericks, P., Grøndahl, L.: Interactions between alginate and chitosan biopolymers characterized using FTIR and XPS. Biomacromolecules **8**(8), 2533–2541 (2007). https://doi.org/10.1021/bm070014y
17. Morán, J.I., Alvarez, V.A., Cyras, V.P., Vázquez, A.: Extraction of cellulose and preparation of nanocellulose from sisal fibers. Cellulose **15**(1), 149–159 (2008). https://doi.org/10.1007/s10570-007-9145-9
18. Lu, P., Hsieh, Y.-L.: Preparation and properties of cellulose nanocrystals: rods, spheres, and network. Carbohydr. Polym. **82**(2), 329–336 (2010)
19. Adsul, M., Soni, S.K., Bhargava, S.K., Bansal, V.: Facile approach for the dispersion of regenerated cellulose in aqueous system in the form of nanoparticles. Biomacromolecules **13**(9), 2890–2895 (2012)
20. Kumar, A., Negi, Y.S., Choudhary, V., Bhardwaj, N.K.: Characterization of cellulose nanocrystals produced by acid-hydrolysis from sugarcane bagasse as agro-waste. J. Mater. Phys. Chem. **2**(1), 1–8 (2014). https://doi.org/10.12691/jmpc-2-1-1

Extraction of Hydrogen Sulfide from Water of Duhok Dam by Industrial Open Pilot Plant

Salah A. Naman and Lazgin A. Jamil

Abstract Hydrogen sulfide is very poisons gas it should be extracted from water and decomposed it to pure sulfur and clean Hydrogen fuel. We had been done the extraction of this gas in Black Sea using two types of pilot plants. This paper will focus on the extraction of low concentration of hydrogen sulfide (H_2S) from Duhok Dam. Survey for the concentration of this gas is low (0.1–5 ppm) at different depth and region of the Dam that may be due to natural sulfur cycle of reducing and oxidizing bacteria in this dam. According to Le Chatelier's principle the equilibrium concentration of H_2S gas is about 0.5 ppm at depth of 15 m. A novel physical extraction method based on Henry's law have been done outside water for stripping H_2S completely from Dam water by special open pilot plant with tower at height between 10 and 15 m depending on the location of plants on the Dam producing cheap and continuous clean water free of H_2S.

Keywords Extraction · H_2S · Duhok dam water · Open pilot plant

1 Introduction

The concentration of H_2S in Duhok Dam is much lower than the concentration of this gas at Black sea water [1, 2]. The production of H_2S in this water is due to the natural sulfur cycles between sulfur reducing bacteria (SRB) and sulfur oxidizing bacteria (SOB). There is a continuous competition reaction of these two types of bacteria and according to the Le Chatelier's principle there is an equilibrium concentration of H_2S in a certain depth of these water lakes [3]. According to the Henry's law, hydrogen sulfide gas in water is non-ideal solution and its extraction will depend on

S. A. Naman (✉) · L. A. Jamil
Department of Chemistry, Faculty of Science, University of Zakho, Zakho, Kurdistan Region, Iraq
e-mail: salah.naman@uoz.edu.krd

L. A. Jamil
e-mail: lazgin.jamil@uoz.edu.krd

Y. T. Mustafa et al. (eds.), *Recent Researches in Earth and Environmental Sciences*, Springer Proceedings in Earth and Environmental Sciences, https://doi.org/10.1007/978-3-030-18641-8_9

the physical properties such as vapor pressure, density, concentration, temperature, pH, and total salt content (TDS) of the solution [4].

Two types of pilot plants have been suggested to increase the efficiency of separation of this gas in Black Sea at pH = 8, concentration of H_2S is 10 ppm and the temperature of 40 °C. Our previous results show that the total concentration of H_2S in aqueous solution is equal to the sum of the concentrations of various sulfide species in solution [5]. These species are independent of pH but the salt-effect would have significant influence on the solubility of H_2S as in Eq. (1).

$$\text{Total } H_2S = m_{H_2S} + mHS^- + mS^= \tag{1}$$

At pH = 7 the concentration of H_2S and $S^=$ are equal and at pH = 8, the concentration of $S^=$ is ten times higher than H_2S [6].

Algebraically, we can express the ideal dilute solution by the following Henry's equation (2).

$$P_j = K_j x_j \tag{2}$$

where the subscript j denotes any of solutes (H_2S) this equation could be for real solution when the solution is dilute and value of K_j for H_2S. This equation relates the partial pressure of H_2S in the vapor phase to the mole fraction of the solute in the solution. Viewing the relation in another way, Henry's law relates the equilibrium mole fraction, the solubility of H_2S in the solution to the partial pressure of H_2S in the vapor equation (3).

$$X_j = (1/K_j)p_j \tag{3}$$

This equation is used to correlate the data on solution of gases in liquids, this facts was discovered by the English physical chemist William Henry (1774–1836). He found that the mass of gas j dissolved by a given volume of solvent at constant temperature is proportional to the pressure of the gas in equilibrium with the solution. Mathematically stated as Eq. (4).

$$m_j = (1/K_j)p_j = kP_j \tag{4}$$

where generally the subscript j is refer to the solute and k is the Henry's constant. Most gases obey Henry's law when the temperatures are not too low and the pressures are moderate [7].

Figure 1 shows application of Henry's Law for SO_2 gas at temperatures of 50 and 70 °C [7].

If several gases from a mixture of gases dissolve in solution, Henry's law applies to each gas independently, regardless of the pressure of the other gases present in the mixture.

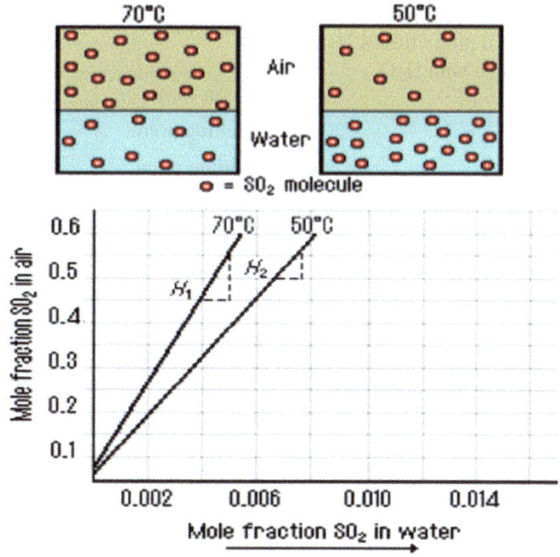

Fig. 1 Application of Henry's law

Although last equation is the historical form of Henry's law, since mass per unit volume is equal to a concentration term we may write this equation as Eq. (5).

$$P_2 = K'x_2 \text{ or } P_2 = K''c_2 \quad (5)$$

The error in determining the exact solubility of vapor pressure of H_2S is multiplied in calculating Henry's Law constant K″. The other error will come due to the effect of temperature on solubility and non-ideality of solution [7]. But the value of $K'' = 515$ (atm) for H_2S. In Chemistry Henry's Law is one of the gas laws formulated by William Henry.

It states, at a constant temperature, the amount of a given gas dissolved in a given type and volume of liquid is directly proportional to the pressure of that gas in equilibrium with that liquid. Figure 1 shows these relations [7]. Also the formulation of Henry's Law is similar to the previous one, and very similar to Arrhenius equation, for dependence of physical and chemical properties of temperatures.

The extraction of H_2S from the Duhok dam will give us pure water suitable for human drinking and pure H_2S. But when H_2S gas is very high in water as in black Sea or in natural gas we could decompose it to pure and cheap sulfur and clean hydrogen fuel [6, 8] because the energy of decomposition of H_2S is much lower than decomposition of water due to bonds structures of both molecules beside the large different in chemical and physical properties for both molecules as in Table 1.

The aim of this research is to apply above principles for separation of poison hydrogen sulfide from water of Duhok dam to be suitable for human drinking after using our laboratory pilot plant by transferring dam water to the laboratory.

Building of new open type of industrial pilot plant at a Duhok dam for local people has been done.

Table 1 The physical and chemical properties of H_2S compared with H_2O

Properties	Hydrogen sulfide (H_2S)	Water (H_2O)
Molar mass	34 g	18 g
Melting point	−85.5 °C	0 °C
Boiling point	−60.7 °C	100 °C
Density	(Vapor) 1.188 g/l at 5 °C	1000 g/l
Solubility in water	3.3 g/l = 3300 ppm 4370 ml/l at 0 °C 1860 ml/l at 40 °C 3.2 g/l at 30 °C	–
Acidity pK	6.89	14
Henry's constant K	515 atm; 919 kPa	–
Heat of fusion H gas	20 kJ/mole	6.013 kJ/mole
Ignition temp.	270 °C	–
Explosion limits	4.3–45.5% V	–

2 Experimental Part

Analysis of H_2S concentration has been done for the dam water at different locations and depths of dam using pump kit and gas detector tube from (Cole-Parmer with different types AN-86513-26, SN 86514-00 and SN 86385-01) as a very sensitive H_2S sensors.

The primary experiment has been studied in our laboratory plant [5] for inlet and outlet water to the plant at different variable (temperature, pumping rate, and tower high), till we get the zero concentrations of H_2S in outlet water.

Then new open industrial pilot plant has been built at the north side of the Duhok dam as shown in Fig. 3. Using the optimized data from our laboratory pilot plant and detection of H_2S in the inlet and outlet water has been measured continuously by H_2S sensors till we get a complete separation of H_2S at outlet tank of water to be suitable for human drinking.

3 Results and Discussion

Duhok dam is nearby Duhok city, Fig. 2. The water of this dam is contaminated with hydrogen sulfide due to Garmava village's spring water. The concentration of H_2S is lower than Black Sea. It's ranging between 0.3 and 5 ppm. The equilibrium concentration of H_2S at depth of 10 m is 1 ppm and this equilibrium concentration changes at different locations of the dam.

Fig. 2 Duhok dam

The water of this dam cannot be used by human drinking and there is missing step for purification of the water (technological process) as the depth of the dam is between 5 and 40 m. The industrial extraction unit could be inside or outside depending on the economic and technical visibility of H_2S separation.

Regarding location and concentration of H_2S in this water for different areas we built two opened pilot plant similar to that of our industrial pilot plant at Black Sea but they are opened on the top with a spray tower at height of 10–20 m as shown in Fig. 3. Separated H_2S will diffused to the atmosphere as it is small.

The open pilot plant working according to the Henry's law containing different sensors for measuring concentration of H_2S, temperature and TDS for raw input water and production clean water after spraying it out to the atmosphere, we are also measuring the H_2S gas on the surface of dam water and also at a top of towers in order not to polluted the atmosphere of the area. These opened pilot plants produced clean water with zero H_2S concentration which it is suitable to human drinking and the price of this pilot plant is cheap depending on the raw water pumping machine and it is made from stainless steel.

The separated H_2S by this system to the atmosphere is not polluting the environment as there are very sensitive H_2S sensors recording the toxicity of H_2S to the human and other living fish [9].

Considering the Environmental and Social Objectives of this project, there are some other classical methods for removing H_2S from the water as the following facts:

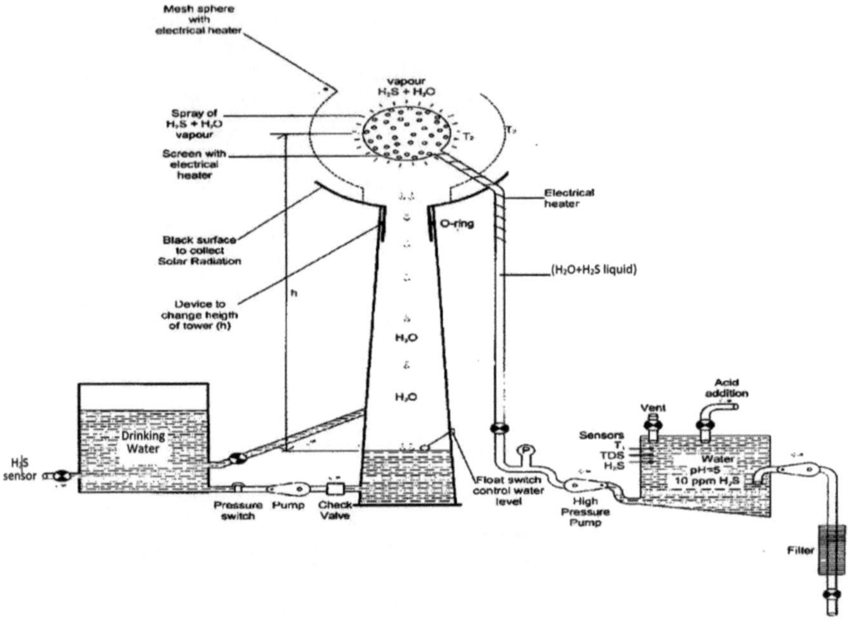

Fig. 3 Opened industrial pilot plant for Duhok dam water

The odor of H_2S gas can be perceived at levels as low as 10 ppb. At levels of 50–100 ppm, it may cause the human sense of smell to fail. Low level can cause eye irritation, dizziness, coughing and headache.

Treatment of H_2S problem depends on its concentration for 1 ppm (mg/l) activated carbon filtration will reduce the unpleasant taste less than 2 ppm. Aeration (adding air to the water) is an appropriate treatment method. In any aeration system, the water must be protected from bacteria contamination and freezing. Limitation of this method is that the aeration of this method is that the aeration process produces a strong H_2S odor near the aeration which may be unpleasant if near the household living area it needs also carbon filter [10].

For H_2S concentration of (1–10 mg/l = 10 ppm), the use of an iron-removal filter containing manganese greensand is supported. For more than 6 ppm constant chlorination using an automatic chemical feed pump is the most common treatment method. The recommended dosage is 2 mg/l = 2 ppm chlorine for each 1 mg/l = 1 ppm H_2S. The chlorine should be added ahead of the mixing tank and sufficient storage must be provided to maintain 20 min of contact line between the water and chemical.

4 Conclusion

1. The concentration of H_2S in Duhok dam water is ranging between 0.1 and 5 ppm and it is not economically visible to collect a hydrogen sulfide and decomposing it to a hydrogen and sulfur that is the reason we built the cheap open industrial pilot plant to remove H_2S at top of the tower to the atmosphere.
2. Survey of concentration of H_2S in different Duhok dam area is low and the equilibrium concentration is different from 0.5 to 1 ppm at depth of 10 m therefore extraction is outside Dam water.
3. This opened pilot plant for extract H_2S will produce to us cheap and continuous water free of H_2S suitable to human drinking.
4. Extraction of H_2S from Duhok dam wills not disturbers the natural equilibrium state of H_2S in a dam according to Le Chatetiers' principle.

References

1. Beauchamp, R., Bus, J.S., Popp, J.A., Boreiko, C.J., Andjelkovich, D.A., Leber, P.: A critical review of the literature on hydrogen sulfide toxicity. CRC Crit. Rev. Toxicol. **13**(1), 25–97 (1984)
2. Midilli, A., Ay, M., Kale, A., Veziroglu, T.N.: Hydrogen energy potential of the Black Sea deep water based on H_2S and importance for the region. In: Proceedings of the International Hydrogen Energy Congress and Exhibition IHEC (2005)
3. Haklıdır, M., Kapkın, Ş.: Black Sea, a hydrogen source. In: Proceedings International Hydrogen Energy Congress and Exhibition IHEC (2005)
4. Dimitrov, D., Dimitrov, P.: Alternative resources and energy sources from the Black Sea bottom. In: 1st Assemblage Workshop, pp. 23–26 (2004)
5. Naman, S., Ture, I.E., Veziroglu, T.N.: Industrial extraction pilot plant for stripping H_2S gas from Black Sea water. Int. J. Hydrogen Energy **33**(22), 6577–6585 (2008)
6. Carroll, J.: Software for Phase Equilibria in Natural Groundwater System. AQV alibrium. http://www3.telus.net/public/jcarroll/ION.HTM
7. Naman, S.A.: Production of Hydrogen from H_2S. Pilot Project of Kirkuk-Iraq, Aug 2006. UNIDO ICHET—International Center for Hydrogen Energy Technologies, Istanbul, Turkey (2006)
8. Al-Shamma, L., Naman, S.: The production and separation of hydrogen and sulfur from thermal decomposition of hydrogen sulfide over vanadium oxide/sulfide catalysts. Int. J. Hydrogen Energy **15**(1), 1–5 (1990)
9. USEPA: http://www.epa.gov/eogapti/module4/absorbtion/absorbtion.html (2010)
10. WHO: Guidelines for Drinking Water Quality Hydrogen Sulfide in Drinking water, vol. WHO/SDE/WSH/03.04/07. WHO, Geneva (1996)

4 Conclusion

References

Modeling the Effect of Reservoir Fluid Properties on Abundance of (H_2S) Evolved from Oil Wells and Dissolved in Reservoir Fluids

Ibtisam Kamal, Keyvan Amjadian, Namam Salih, Bryar Ahmad and Rebwar Haidar

Abstract Oil reservoirs are generally described by its main characteristics including its lithology and the pressure-volume-temperature (PVT) data. In reservoir engineering calculations, PVT data are essential to determine the reservoir fluid composition and design the production facilities. PVT data of 10 petroleum crude wells of Tawke field—Zakho are investigated in the current study. The data were analyzed to define the mathematical models correlate and govern the reservoir fluids properties as a function of sulfur compounds in term of (H_2S) evolved from the wells and dissolved in reservoir fluids. The results obtained showed that crude oils are classified as sour, and medium with high content of heavier fractions C6+ up to C14+ (average ^{o}API 23.4). The mathematical models explain the correlations were estimated. H_2S evolved and dissolved in reservoir fluids proved to increase with increasing sample depth, reservoir pressure and temperature, gas oil ratio, coefficient of compressibility at reservoir pressure, and decreases with increasing ^{o}API, viscosity at reservoir pressure and bubble point pressure, and gross heating value.

Keywords PVT data · Tawke field · Hydrogen sulfide · Mathematical modeling

I. Kamal (✉)
Chemical Engineering Department, Faculty of Engineering, Soran University, Soran, Kurdistan Region, Iraq
e-mail: Ibtisam.kamal@soran.edu.iq

K. Amjadian · N. Salih · B. Ahmad · R. Haidar
Petroleum Geosciences Department, Faculty of Engineering, Soran University, Soran, Kurdistan Region, Iraq
e-mail: keyvan.amjadian@soran.edu.iq

N. Salih
e-mail: namam.muhammed@gmail.com

B. Ahmad
e-mail: bryar.ahmad1985@gmail.com

R. Haidar
e-mail: rebwar.haidar95@gmail.com

Y. T. Mustafa et al. (eds.), *Recent Researches in Earth and Environmental Sciences*, Springer Proceedings in Earth and Environmental Sciences,
https://doi.org/10.1007/978-3-030-18641-8_10

1 Introduction

Sulfur is the tenth most numerous element in the universe. In nature, sulfur can primarily be found occurring in three forms; elemental sulfur, sulfides, and sulfates. Sediments accumulated about 43% of the total sulfur of the earth's crust [1].

The decomposition of organic matter of crude oil results in formation of Sulfur. The aqueous sulfate derived from water, or from dissolution of solid calcium sulfate can be reduced at high temperature (above 250 °C) by a variety of organic compounds such as alcohols, polar aromatic and saturated hydrocarbons via a thermal redox reactions to produce high concentrations of H_2S in the reservoir fluid (more than 10%) [2]. the phenomena have been predicted from the data of sulfur isotopes [3]. The produced H_2S appears in the associated gas, and some portion of sulfur stays with the liquid. H_2S can react with the hydrocarbons to give sulfur compounds based on pressure, temperature and period of formation of the reservoir [4]. Furthermore, temperature, pressure, fluids composition, and water pH and ionic strength are the factors that control the thermodynamic process of partitioning of H_2S between the oil, water and gas [5]. H_2S is so much of a risk that sour crude has to be stabilized via removal of H_2S before it can be transported by oil tankers. H_2S gas is flammable and extremely toxic, it can be fatal when its concentration is as low as 100 ppm [6]. An oil well is deemed to be sour if the concentration of H_2S exceeds 10 ppm in the gas phase. Because of H_2S is toxic and causes corrosion, plugging of reservoir formations, and increasing sulfur content of the produced oil, precautions are necessary to be taken in design and operation of production, transport, and storage equipment [7–9]. Also, sulfur content increases with densification of crude oil [10, 11].

On the other hand, it is well established that reservoir fluids characteristics are known to vary with pressure, temperature, and volume of the system. All reservoir engineering calculations require PVT data to determine the reservoir fluid composition and design the production facilities [12, 13]. Because sulfur content is one of the major factors that affect crude oils quality and price, estimation the correlations between PVT data and sulfur content are very important. However, it is complicated to obtain universal correlations between PVT data and sulfur content. As alternative, correlations for local regions, where crude properties are expected to be uniform are acceptable.

Studies on predictions the correlations of sulfur compounds content with PVT data for crude oils of Kurdistan Region—Iraq are scarce. The objective of the current work is to predict the mathematical models that govern the correlation between sulfur compounds content based on H_2S and reservoir characteristics. Tawke field is an oil field located close to Iraq's border with Turkey in the semi-autonomous Kurdistan Region, has been named as the oil field of the theme of the study. The reservoir characteristics were analyzed based on PVT data supported by DNO [14]. The lithology of the area has been described, and concentration of H_2S evolved from the oil wells and that dissolved in reservoir fluids was correlated with reservoir characteristics. The mathematical correlations were estimated.

2 Methodology: Lithology of the Petroleum Wells and the Reservoir Fluid Properties

The topography of the area of the oil wells under investigation was inspected. The findings revealed the presence of folds and faults, which are widely spread, this had significant role in shaping the complex topography characterized by high altitude and very steep slopes composing cliffs of carbonate rocks. In this location, the relatively long and narrow main valley is surrounded by several high mountains Bekher Mountain, at the southeastern of Tawke village, and Shiranish Islam to the north.

The stratigraphic units of Miocene observed (mostly in subsurface sections of Tawke wells) are: (1) Fatha (Lower Fars) Formation, (2) Jeribe Formation, (3) Dhiban Formation, (4) Euphrate Formation, (5) Pila Spi Formation.

The main lithology of the wells is including sandstone, claystone, anhydrite, siltstone, limestone, dolomite, marl and shale which are due to the geology of the area where wells are drilled within geological units called Lower Fars, Jeribe, Dhiban, Euphrates and Pila Spi formations. The lithology of the ten investigated wells is described briefly below.

Well A: It is a 313 m depth well which is mainly consists of sandstone, claystone, anhydrite, limestone and dolomite within Lower Fars and Jeribe formations. Well B: A 298 m depth well including sandstone, claystone, siltstone, anhydrite, limestone in Lower Fars and Jeribe formations. Well C: A well of 405 m depth with sandstone, claystone, anhydrite, siltstone and limestone in Lower Fars and Jeribe formations. Well D: It is a 307 m depth well consisting of sandstone, claystone, anhydrite, siltstone and limestone in formations of Lower Fars and Jeribe. Well E: A 636 m depth well drilled in Lower Fars, Jeribe, Dhiban, Euphrates and Pila Spi formations showing sandstone, claystone, siltstone, limestone, marl, anhydrite and dolomite. Well F: It is consisting of claystone, sandstone, siltstone, anhydrite, marl, limestone and dolomite within Lower Fars, Jeribe, Dhiban, Euphrates and Pila Spi formations in a 730 m drilled well. Well G: A well of 580 m depth with sandstone, claystone, siltstone, anhydrite, limestone and dolomite drilled in Lower Fars, Jeribe, Dhiban and Euphrates formations. Well H: 395 m depth well drilled in Lower Fars, Jeribe and Dhiban formations including claystone, sandstone, siltstone, anhydrite, limestone and shale. Well I: A well of 712 m depth drilled in Lower Fars, Jeribe, Dhiban, Euphrates and Pila Spi formations with claystone, sandstone, siltstone, anhydrite and limestone. Well J: It is consisting of sandstone, claystone, anhydrite and limestone within Lower Fars, Jeribe, Dhiban, Euphrates and Pila Spi formations in a 725 m drilled well.

In summary, the main lithology of the wells is including sandstone, claystone, anhydrite, siltstone, limestone, dolomite, marl and shale. The depth of the wells range from 298 to 725 m. A typical master log is presented in Fig. 1 (well A).

As known, reservoir fluids could be classified to three categories; (i) aqueous solutions with dissolved salts, (ii) liquid hydrocarbons, and (iii) gases (hydrocarbon and non-hydrocarbon). The compositions of reservoir fluids depend on their source, history, and present thermodynamic conditions. The distribution of reservoir fluids

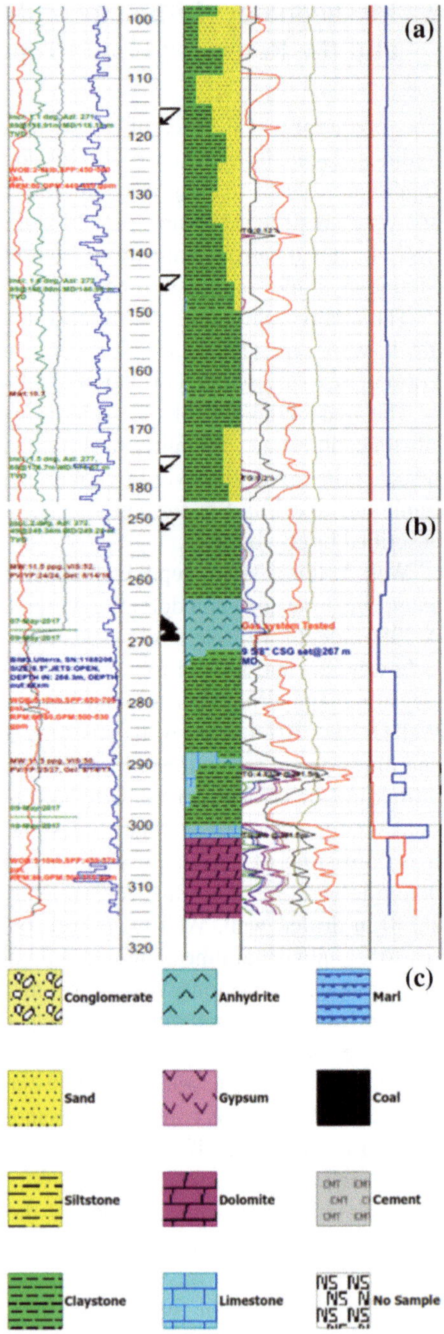

Fig. 1 A typical master log and lithostratigraphy (well A)

within a given reservoir depends upon the physical and chemical properties of the fluids, the thermodynamic conditions of the reservoir as well as the petro-physical properties of the rocks. Gravity, capillary, molecular diffusion, thermal convection, and pressure gradients are the forces that originally distribute the fluids.

Reservoir fluid properties has been highlighted and overviewed in literature [15–17]. However, studies on predictions the correlations of sulfur compounds content with reservoir characteristics including the PVT data for crude oils are very scarce. In the current work, several reservoir fluids parameters were correlated with the amount of Hydrogen Sulfide (H_2S) evolved from the reservoirs and dissolved in reservoir fluids for 10 petroleum wells distributed in Tawke oil field. The reservoir fluid properties correlated include:

2.1 Gas Gravity

Gas gravity could be defined as the molecular weight of the gas divided by the molar mass of air (28.94 kg/kmole). Petroleum gases typically have a gravity of about 0.65. The calculations of solution gas-oil-ratio, gas viscosity, compressibility as well as compressibility factor are highly affected by gas gravity.

2.2 Solution Gas-Oil Ratio (GOR)

The solution gas-oil ratio is the amount of gas dissolved in the oil at any pressure. The GOR is directly proportional approximately with pressure. It increases with pressure until the bubble point pressure is reached, after which it becomes constant, and at that time the oil is said to be under saturated. Also, the GOR is a function of the oil and gas composition, light oils contain more dissolved gas than heavy oils. In general, the GOR ranges from 0 for dead oil to approximately 2000 scf/bbl for very light oil.

2.3 Reservoir Temperature

Reservoir temperature is the average temperature maintained inside a hydrocarbon reservoir.

2.4 Reservoir Pressure

Reservoir pressure, is the pressure of the fluids present in a hydrocarbon reservoir. It is also known as formation pressure or hydrostatic pressure It can also defined as the pressure which is exerted by column of water on sea level from the depth of a hydrocarbon reservoir inside the earth's surface.

2.5 API Specific Gravity

Specific gravity is one of the significant oil properties. $^{\circ}API = (141.5/\gamma) - 131.5$, γ is the oil specific gravity of the petroleum liquid being measured.

2.6 Reservoir Fluid Density at Bubble Pressure

The bubble point pressure is stated as the pressure at which the first bubble of gas comes out of solution. At this point, the oil is saturated; it cannot hold anymore gas. Above bubble point pressure the oil acts as a single-phase liquid. Lowering of the pressure at and below bubble point pressure leads to liberation of gas resulting in two-phase flow.

2.7 Oil Viscosity

Oil viscosity is a measure of the flow resistance of the oil, and is given in units of centipoises (cP). Higher values of oil viscosity indicate greater resistance to flow. Increasing temperature and pressure (up to the bubble point) results in decreasing the oil viscosity. Oil viscosity increases minimally with increasing pressure (above the bubble point pressure). Reservoir temperature, oil gravity, and solution gas-oil ratio are highly related to oil viscosity.

2.8 Coefficient of Compressibility at Reservoir Pressure

The isothermal oil compressibility is defined as the unit change of volume with pressure. Reservoir calculations and accuracy of design of high-pressure surface equipment are dependable on the isothermal oil compressibility. Higher accuracy of oil compressibility estimates will improve the design of high-pressure surface equipment and material balance calculations. The oil compressibility can be calculated from the slope of the curve of isotherm of the logarithm of the oil formation volume factor versus pressure [18]. Also, it can be determined from the differentiation of an equation of state [19].

3 Results and Discussion

Figures 2, 3, 4, 5, 6, 7, 8, 9 and 10 illustrate the correlations estimated in the current study. Figure 2 shows that as sample depth increases the amount of H_2S evolved and dissolved in reservoir fluids increases. More depth means that the sample is exposed to higher temperature and pressure, resulted in increasing the decomposition rate of sulfur compounds and consequently releasing more H_2S [20, 21] as reflected in Fig. 3.

Figure 4 shows that increasing the temperature of reservoirs resulted in increase the amount of H_2S. The situation is attributed to increasing the rate of degradation

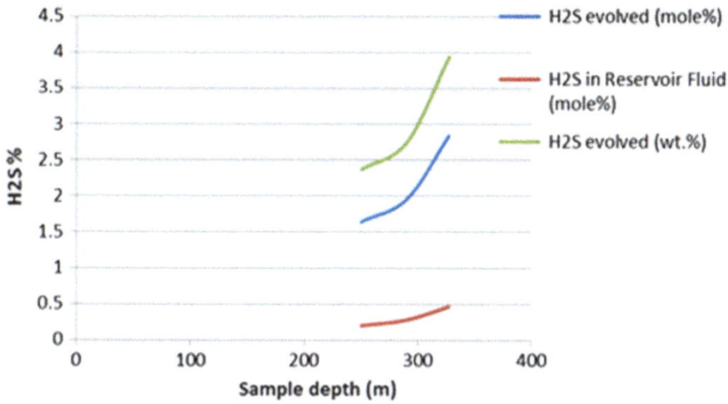

Fig. 2 Amount of H_2S evolved and dissolved in reservoir fluids versus sample depth

Fig. 3 Amount of H_2S evolved and dissolved versus reservoir pressure

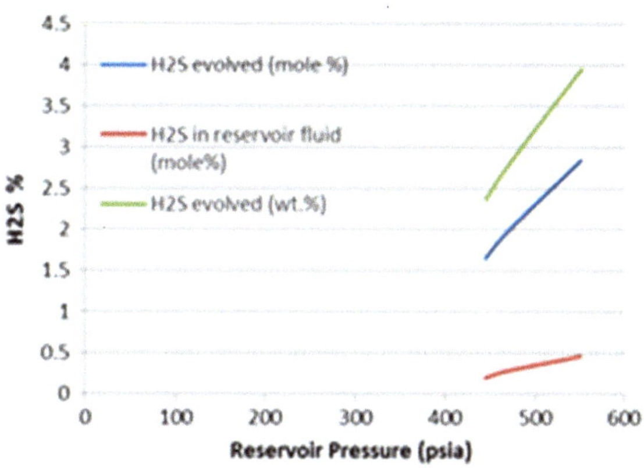

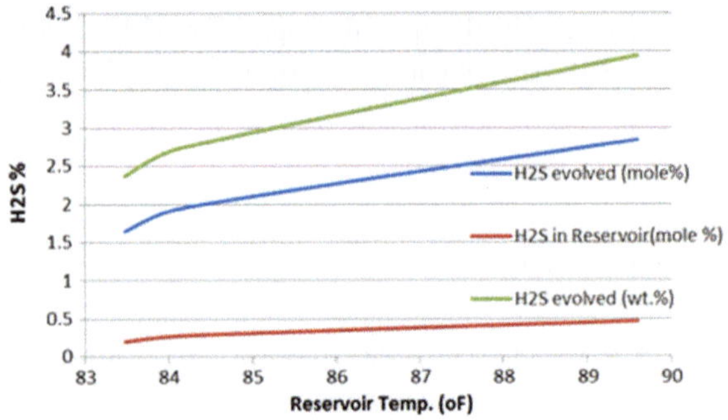

Fig. 4 Amount of H_2S evolved and dissolved versus reservoir temperature

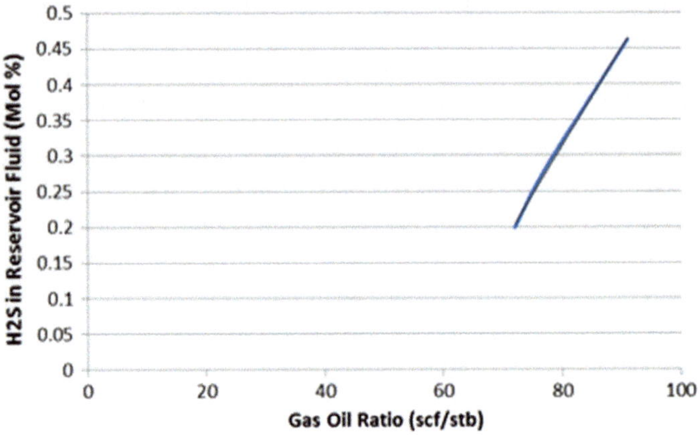

Fig. 5 The amount of H_2S in reservoir fluids as a function of gas oil ratio

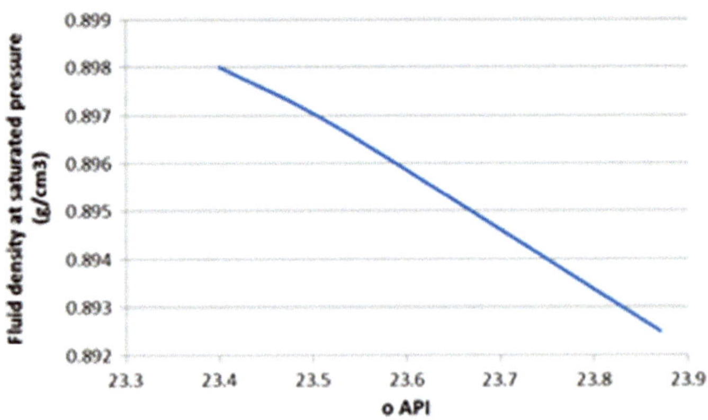

Fig. 6 The variation of fluid density with °API

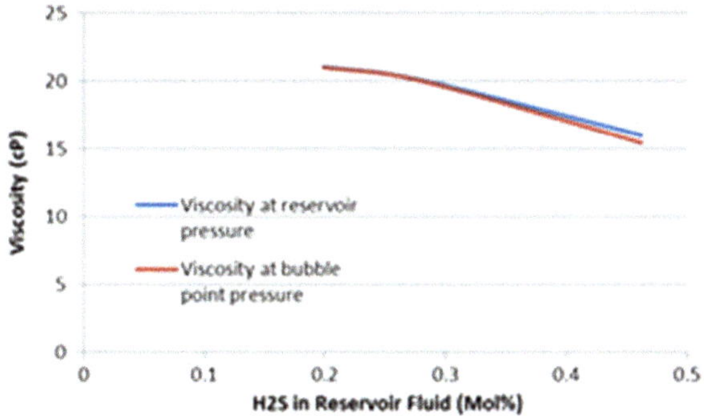

Fig. 7 The amount of H_2S as a function of reservoir fluid viscosity

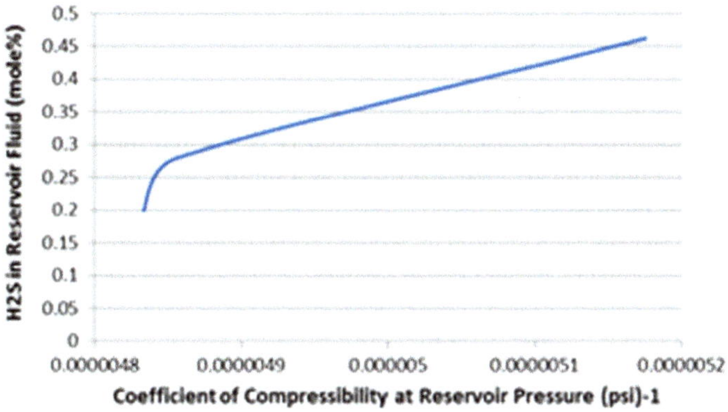

Fig. 8 The variation of the amount of H_2S with coefficient of compressibility

Fig. 9 Evolved gas gravity versus sample with depth

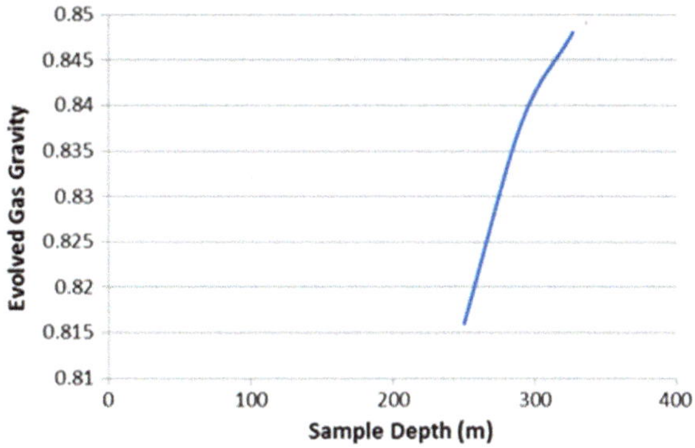

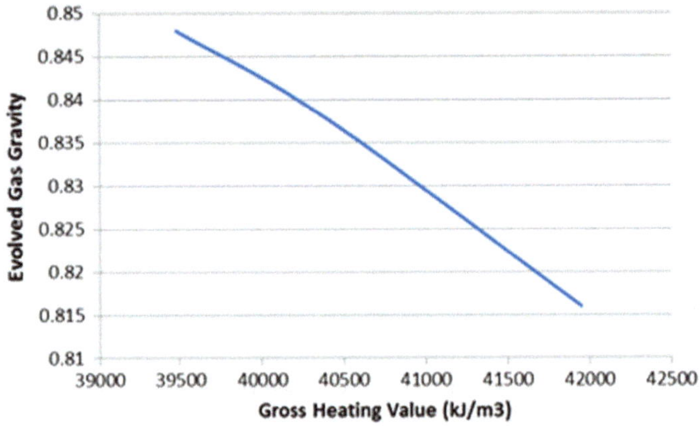

Fig. 10 Evolved gas gravity versus HHV

of sulfur compounds to release more evolved H_2S gas and more dissolved gas in reservoir fluids enhancing by increasing the gas solubility as temperature and pressure of the reservoir increases.

Figure 5 shows how the amount of H_2S in reservoir fluids increases as gas oil ratio increases. It is logical to estimate such relation as gas oil ratio reflects the amount of dissolved gas at any pressure. Figure 6 shows the relation of fluid density with °API. In general light oils are of high °API and lower density compared to heavy oils.

On another hand, the design of transport equipment, such as oil pipelines or compressors, simulation of oil and gas reservoirs production profiles, enhanced oil recovery, and the storage of natural gas are among the engineering disciplines that are highly correlated with viscosity. The amount of H_2S evolved and dissolved in reservoir fluids seemed to decrease with increasing the viscosity at reservoir pressure and at bubble point pressure as shown in Fig. 7.

Figure 8 shows the variation of the amount of H_2S dissolved in reservoir fluid with coefficient of compressibility at reservoir pressure. The results estimated confirmed that the amount of H_2S dissolved in reservoir fluid has a significant positive effect on the unit change of reservoir fluid volume with pressure. As mentioned before, the design of high-pressure surface equipment and reservoir calculations are strongly affected by the isothermal oil compressibility [19].

Figure 9 illustrates the relation between the gravity of the evolved gas from reservoir (including H_2S) with sample depth. Our results demonstrated that as the sample depth increases the gas gravity increases significantly. The results obtained revealed that at higher depths the amount of heavier components including the degradable sulfur compounds capable to release H_2S gas is more than that in samples of lower depths.

Figure 10 demonstrates the variation of evolved gas (including H_2S) gravity versus gross heating value (HHV). It is well known that the energy content of petroleum and its related fuels is defined by HHV. It was pointed out that there is a significant

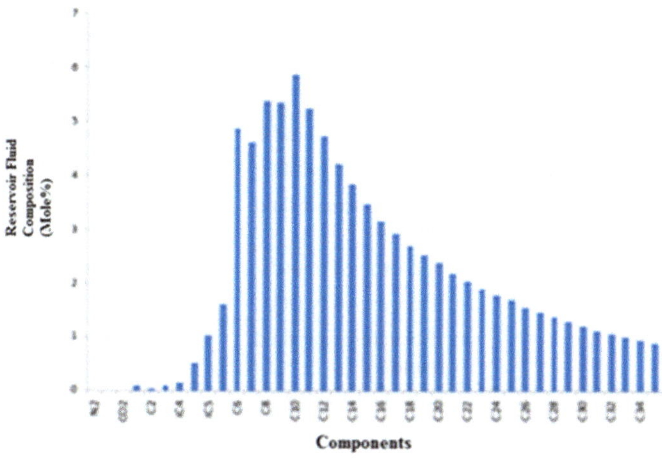

Fig. 11 Typical reservoir fluid composition

correlation between specific gravity and HHVs for petroleum fuels. Regular decrease in HHVs is noted with increasing the specific gravity of petroleum fuels [22]. Our results revealed that HHV of the reservoir fluids increases with increasing the gravity of evolved gas (including H_2S). The results reflect consequently that as the amount of evolved gas increases, the reservoir fluids will contain less dissolved H_2S resulted in increasing the HHV of the reservoir fluids, thus, removing H_2S from reservoir fluids is essential to increase the HHV. This phenomenon is very important from economical point of view.

On the other hand, general estimated characteristics of the investigated back oils from the 10 investigated wells pointed out that the crude oils are classified as medium (average degree API is 23.4) with high content of heavier fractions C6+ up to C14+. Also, they are sour (H_2S content is relatively high) the evolved and the dissolved in reservoir fluids. Figure 11 demonstrates a typical diagram of the reservoir fluid composition from pressure-volume-temperature (PVT) analysis.

The mathematical models derived from the correlations plotted in Figs. 2, 3, 4, 5, 6, 7, 8, 9, 10 and 11 are listed and highlighted in Table 1.

It is obvious to note that Table 1 includes 13 adopted correlations with their mathematical models. It is well remarked that all the models have high regression coefficient reflecting a good fitting of the experimental data. The mathematical models can serve to predict the amount of H_2S evolved and dissolved in reservoir fluids based on reservoir characteristics.

Table 1 Reservoir properties and the estimated mathematical models

	Correlated parameters		Mathematical model	(R^2)
	y	x		
1	Sample depth (m)	H_2S evolved (mole%)	y = 0.0111x − 3.249	0.998
		H_2S evolved (wt%)	y = 0.0111x − 3.249	0.998
		H_2S in reservoir fluid (mole%)	y = 0.0024x − 0.8591	0.991
2	Reservoir temp. (°F)	H_2S evolved (mole%)	y = 0.1829x − 13.539	0.978
		H_2S evolved (wt%)	y = 0.243x − 17.812	0.984
		H_2S in reservoir fluid (mole%)	y = 0.0396x − 3.0788	0.963
3	Gas oil ratio (scf/stb)	H_2S in reservoir fluid (mole%)	y = 0.0137x − 0.7807	0.998
4	Viscosity (cP) at reservoir pressure	H_2S in reservoir fluid (mole%)	y = −19.796x + 25.249	0.985
5	Viscosity (cP) at bubble point pressure	H_2S in reservoir fluid (mole%)	y = −21.79x + 25.648	0.981
6	Coefficient of compressibility at reservoir pressure Psi^{-1}	H_2S in reservoir fluid (mole%)	y = 683,500x − 3.073	0.942
7	Sample depth (m)	Evolved gas (including H_2S) gravity	y = 0.0004x + 0.7134	0.964
S	Gross heating value (kJ/m^3)	Evolved gas (including H_2S) gravity	y = −1E−05x + 1.3541	0.998
9	°API	Fluid density at saturated pressure (g/cm^3)	y = −0.0118x + 1.175	0.998

4 Conclusions

The investigated data revealed that the crude oils are classified as sour, and medium (average degree API is 23.4) with high content of heavier fractions C6+ up to C14+. Thirteen adopted mathematical models describe the effect of reservoir properties on amount of H_2S evolved from oil wells and dissolved in reservoir fluids were estimated. All the models were of high regression coefficients assuring a high fitting of the experimental data. The mount of sulfur compounds seemed to increase with increasing sample depth, reservoir pressure and temperature, gas oil ratio, coefficient of compressibility at reservoir pressure, while °API, viscosity and gross heating value of reservoir fluid proved to decrease with increasing the H_2S evolved from oil wells and dissolved in reservoir fluids.

Acknowledgements This work is developed by a partnership between Soran University and Ministry of Natural Resources of Kurdistan Region under the thematic network of research and engineering. The authors wish to acknowledge the technical staff of DNO; the Norwegian oil and gas operator in Kurdistan Region, for supporting the data analyzed in this investigation.

References

1. Wedepohl, K.H.: Sulfur in the earth's crust, its origin and natural cycle. Stud. Inorg. Chem. **5**, 39–54 (1984)
2. Basafa, M., Hawboldt, K.: Reservoir souring: sulfur chemistry in offshore oil and gas reservoir fluids. J. Petrol. Explor. Prod. Technol. **9**(2), 1105–1118 (2019)
3. Zhu, G., Liu, X., Yang, H., Su, J., Zhu, Y., Wang, Y., Sun, C.: Genesis and distribution of hydrogen sulfide in deep heavy oil of the Halahatang area in the Tarim basin, China. J. Nat. Gas Geosci. **2**(1), 57–71 (2017)
4. Wauquier, J.-P.: Petroleum Refining: Crude Oil, Petroleum Products, Process Flowsheets, vol. 1. Editions Technip (1995)
5. Burger, E.D., Jenneman, G.E., Carroll, J.J.: On the partitioning of hydrogen sulfide in oilfield systems. In: SPE International Symposium on Oilfield Chemistry. Society of Petroleum Engineers (2013)
6. Heshka, N.E., Hager, D.B.: Measurement of H_2S in crude oil and crude oil headspace using multidimensional gas chromatography, deans switching and sulfur-selective detection. J. Vis. Exp. (106), 53416 (2015)
7. Elshiekh, T., Elmawgoud, H., Khalil, S., Alsabagh, A.: Simulation for estimation of hydrogen sulfide scavenger injection dose rate for treatment of crude oil. Egypt. J. Pet. **24**(4), 469–474 (2015)
8. Holubnyak, Y., Bremer, J.M., Hamling, J.A., Huffman, B.L., Mibeck, B., Klapperich, R.J., Smith, S.A., Sorensen, J.A., Harju, J.A.: Understanding the souring at Bakken oil reservoirs. In: SPE International Symposium on Oilfield Chemistry. Society of Petroleum Engineers (2011)
9. Eden, B., Laycock, P., Fielder, M.: Oilfield reservoir souring. Health and Safety Executive Offshore Technology Report, OTH 92 385. HSE Books, Sudbury, Suffolk, UK (1993)
10. Demirbas, A., Alidrisi, H., Balubaid, M.: API gravity, sulfur content, and desulfurization of crude oil. Pet. Sci. Technol. **33**(1), 93–101 (2015)
11. Awadh, S.M., Al-Mimar, H.: Statistical analysis of the relations between API, specific gravity and sulfur content in the universal crude oil. Int. J. Sci. Res. **4**(5), 1279–1284 (2015)
12. Dake, L.P.: Fundamentals of Reservoir Engineering, vol. 8. Elsevier (1983)
13. Sylvester, O., Samuel, O., Bibobra, I.: PVT analysis reports of Akpet GT9 and GT12 reservoirs. Am. J. Manage. Sci. Eng. **2**, 132–144 (2017)
14. Black oil PVT analysis, Emirates link Weatherford Laboratories Job file Number: F1713, DNO, Iraq.
15. Bateman, R.M.: Cased-Hole Log Analysis and Reservoir Performance Monitoring. Springer (1985)
16. Ahmed, T.: Reservoir Engineering Handbook. Gulf Professional Publishing (2018)
17. Fanchi, J.R., Christiansen, R.L.: Introduction to Petroleum Engineering. Wiley (2016)
18. Al-Marhoun, M.A.: The coefficient of isothermal compressibility of black oils. In: Middle East Oil Show. Society of Petroleum Engineers (2003)
19. Adepoju, O.O.: Coefficient of Isothermal Oil Compressibility for Reservoir Fluids by Cubic Equation-of-State (2006)
20. Xu, C., Zou, W., Yang, Y., Duan, Y., Shen, Y., Luo, B., Ni, C., Fu, X., Zhang, J.: Status and prospects of deep oil and gas resources exploration and development onshore China. J. Nat. Gas Geosci. **3**(1), 11–24 (2018)

21. Wenhui, L., Bo, G., Zhongning, Z., Jianyong, Z., Dianwei, Z., Ming, F., Xiaodong, F., Lunju, Z., Quanyou, L.: H_2S formation and enrichment mechanisms in medium to large scale natural gas fields (reservoirs) in the Sichuan Basin. Pet. Explor. Dev. **37**(5), 513–522 (2010)
22. Demirbas, A., Al-Ghamdi, K.: Relationships between specific gravities and higher heating values of petroleum components. Pet. Sci. Technol. **33**(6), 732–740 (2015)

Hydrocarbon Degradation of Oil Pipeline Blockage by Thermophilic Fungi Isolated from Tawke Field

Yousif A. AlBany, Anwer N. Mamdoh and Mohammad I. Al-Berfkani

Abstract Recently major challenges facing oil industrials due to blockage of oil pipelines. Bioremediation technique was used for clearing the blockage of oil pipelines. Soil samples contaminated with Crude oil was collected from Tawke field in the Kurdistan Region of Iraq. Different strains of fungus were isolated from soil and estimated their hydrocarbon degradation efficacy. The fungal efficient for oil degradation was characterized by study their morphology and estimate their substrate utilization. The effect of following parameters such as Temperature, pH, the concentration of oil and NaCl on the efficiency of fungus was studied. The result found that *Penicillium expansum* has ability to degrade the crude oil at 10% concentration associated with 0.5% NaCl under microaerophilic environment in one week at 55 °C at pH 7. The activity of biodegradation was confirmed by using Thin Layer Chromatography (TLC). It is one of the major advantage using *Penicillium expansum* in large scale production as these strains considered thermophilic strains which have merit over using Mesophilic strain due to less cost for cooling during production and minimize the contamination.

Keywords Oil pipeline blockage · Hydrocarbon degradation · Fungi · Thermophilic

Y. A. AlBany (✉)
Nursing Department, Bardarash Technical Institute, Duhok Polytechnic University, Duhok, Kurdistan Region, Iraq
e-mail: yousif.albany@yahoo.com

A. N. Mamdoh
Pharmacology Department, Duhok Technical Institute, Duhok Polytechnic University, Duhok, Kurdistan Region, Iraq
e-mail: answed80@gmail.com

M. I. Al-Berfkani
Medical Laboratory Techniques Department, Zakho Technical Institute, Duhok Polytechnic University, Duhok, Kurdistan Region, Iraq
e-mail: mohammad.said@dpu.edu.krd

Y. T. Mustafa et al. (eds.), *Recent Researches in Earth and Environmental Sciences*,
Springer Proceedings in Earth and Environmental Sciences,
https://doi.org/10.1007/978-3-030-18641-8_11

1 Introduction

Many substances deposited in the pipeline and as consequence block the oil pipeline and that based on the chemical characterizes of the liquid transport within the pipeline. Such substances that cause blocking the pipeline are wax and paraffin as heavy hydrocarbons or asphaltene as organic material [1]. Once the substance deposits in the pipeline it hindered and delayed the flow of the fuel and hence increased the pressure inside the pipeline and later caused the corrosion and weakness of the pipeline walls. Although two standard methods are used to remove the block in the pipeline; Mechanical and chemical cleaning methods, these two methods facing many problems. Hence scientific researchers aim to use biological ways to treat deposition problems [2].

The oil has heavy hydrocarbons and high initial viscosity and the oil density is 0.9753 kg/dm^3, which is near the density of water (1 kg/dm^3). The insolubility of oil in water is due to little mixing occurs between the oil and water hence no formation of suspension or solution. Although oil has low volatility in atmosphere and not evaporate, some oil has a very gaseous composition and if spilled, will evaporate [3, 4]. All of these characteristics contribute to the difficulty with clean-up methods. Therefore, the oil needs to be cleaned up via human interaction, or degradation. The oxidation of hydrocarbon biologically first observed in last century by bacteria and fungus [5, 6]. Saturated hydrocarbons are triggered aerobically by monooxygenase reaction. Both enzymes monooxygenase and dioxygenase produced by microorganisms attach aromatic hydrocarbons [7].

One of the most form of biodegradation is Mycoremediation, by using fungus in the process of convert contaminated environment with pollutants to less state. Paul Stamets was first who invented the term Mycoremediation to describe the use of microorganisms such as fungus in the biodegradation. The ecosystem decomposition is rely on the mycelium fungus which produced many acids and extracellular enzymes lead to break down long chain of Carbon and hydrogen. It is important to select the right species of fungus to target specific pollutant [8, 9]. Petroleum hydrocarbons have been biodegraded in both terrestrial and marine ecosystem. Hydrocarbons biodegradation by thermophilic microorganisms with low water solubility is now interested as at high temperature both solubility and bioavailability enhanced [10]. Hence the field of used microorganisms for hydrocarbon degradation is developed, and may further be used for decontamination processes [11]. Bacteria and filamentous fungi are the main potential agents used for consuming hydrocarbon petroleum. Once these microorganisms grow on the substrate the start produced different enzymes for degradation [12–14]. Use of fungus as biodegradation agent is more effective due to their ability to grow under stress condition such as low water availability, low pH and even in case of poor nutrients. Many research studies reported that filamentous fungus could be used as bioremediation microorganisms for treatment of pollution [13, 15–17]. The aim of present study is to identify the fungus species which have ability solve the problem of oil pipeline blockage by their capability to degrade petroleum hydrocarbons. Hence, this study is finding the best hydrocarbon degrading fungi and checking its degradation ability at different parameters like temperature, pH, concentration of oil, concentration of sodium chloride.

2 Materials and Methods

2.1 Chemicals and Media

1. The oil and oil contaminated soil sample used in the study was from Tawke field, Zakho, Duhok.
2. Media used for fungal cultivation was Potato Dextrose Agar (Himedia).
3. The Thin Layer Chromatography (TLC) plates were from Merck Ltd. The solvent system used for TLC was ethyl acetate + n-hexane (1:1).
4. Standard oils—Pentadecane, Nonadecane and Heavy Liquid Paraffin from SD fine Company.

2.2 Isolation and Enrichment of Hydrocarbon Degrading Fungi

The fungal strains were isolated from the oil contaminated soil sample from Tawke field. For isolation, mineral medium plates with 1 mL crude oil sample were used which were inoculated with 1 g of soil sample (106 diluted) [Composition of mineral media (per 1 L) Solution A (0.9 L)—K_2HPO_4 (0.8 g), KH_2PO_4 (0.2 g), KCl (0.1 g), $Na_2MoO_4.2H_2O$ (0.025 g), Na_2Fe EDTA (0.014 g), NH_4NO_3 (1 g). Solution B (0.1 L)—$MgSO_4.7H_2O$ (2 g), $CaCl_2.H_2O$ (0.1 g), Agar (30 g)] [18]. These plates incubated at 55 °C for 8 days (pH 7). The obtained fungus was purified on Potato Dextrose Agar. The isolates were routinely sub cultured and stored [18].

2.3 Characterization of Fungal Strains

The characterization of obtained fungal isolates was done by microscopic observations using Lactophenol stain [19].

2.4 *Degradation of Hydrocarbon by Isolated Strain*

The hydrocarbon degradation by Fungi was monitored in Flasks containing 100 mL mineral medium (composition as above mentioned) with 1 mL oil sample and incubated at 55 °C for 8 days on rotary shaker incubator at 70 rpm.

The degradation activity was determined by checking absorbance at 700 nm and by TLC.

Optimization: The degradation efficiency of the best degrader was assessed at various different parameters like

1. pH (4, 5, 7, 8)
2. Temperature (5, 30, 45, 55, 60 °C)
3. Concentration of oil (10, 15, 20%)
4. Concentration of NaCl (0.5, 1, 4, 6%)
5. Carbon number (Pentadecane-15C, Nonadecane-19C, Heavy liquid Paraffin (1622C).

3 Results

3.1 *Isolation of Hydrocarbon Degrading Fungi*

8 isolates were obtained by enrichment of soil sample in mineral medium in 8 days incubation at 55 °C (Fig. 1).

Morphological characterization of oil degrading fungi:

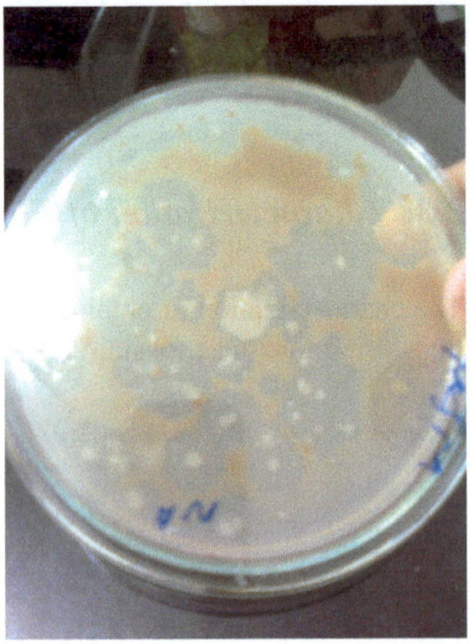

Fig. 1 Growth of bacteria and fungi on oil contaminated soil

Table 1 Morphological characteristics of fungal isolates

Fungal isolate	Macroscopic characteristics	Microscopic characteristics
F_A	Colonies fast growing, fluffy, cotton candy appearance	Hyphae—non-septate. Sporangia—round, black, filled with sporangia
F_B	Colonies—green, velvety appearance	Mycelial hyphae—septate. Conidia—ellipsoidal
F_C	Colonies—white, velvety appearance	Mycelial hyphae—septate. Conidia—ellipsoidal
F_D	Colonies—blue green, velvety appearance	Mycelial hyphae—septate. Conidia—ellipsoidal
F_E	Colonies grow rapidly, woolly to cottony flat appearance, white in color	The cushion like mat of hyphae bearing conidiophores usually absent in culture
F_F	Colonies rapidly growing, fluffy and orange in appearance	Prostrate and aerial hyphae, granules scattered in sectors
F_G	Colonies—blue green, velvety appearance	Mycelial hyphae—septate. Conidia—ellipsoidal
F_H	Colonies grow rapidly, initially white and then turns black. Conidial heads are carbon black	Mycelial hyphae—septate and colorless. Conidia—spherical contains conidiophores

Most of the fungi were filamentous and spore forming when observed with Lactophenol staining as shown in Table 1.

3.2 *Estimation of Degrading Ability*

Monitoring the degrading activity of each isolate based on Absorbance at 700 nm after 8 days incubation at 55 °C at 70 rpm as shown in Fig. 2.

The graph reveals that the fungal isolate FG, showed the lowest absorbance after 8 days of inoculation at 55 °C and pH 7. The morphological characterization suggested that the fungal isolate FG belongs to the genus Penicillium.

3.3 *Effect of Different Parameters Degradation Activity of Penicillium*

Figure 3, showed the effect of pH on the ability of fungus to degrade the oil, it showed that maximum degradation activity was observed at pH 7, in which the absorbance was 0.03 at 700 nm. Which explain the neutral environment at pH 7 is the best condition for fungus degradation.

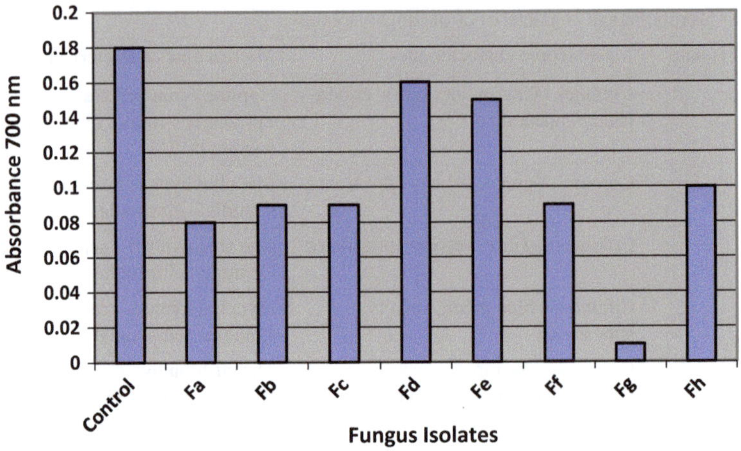

Fig. 2 Estimation of degrading ability of each fungal isolate

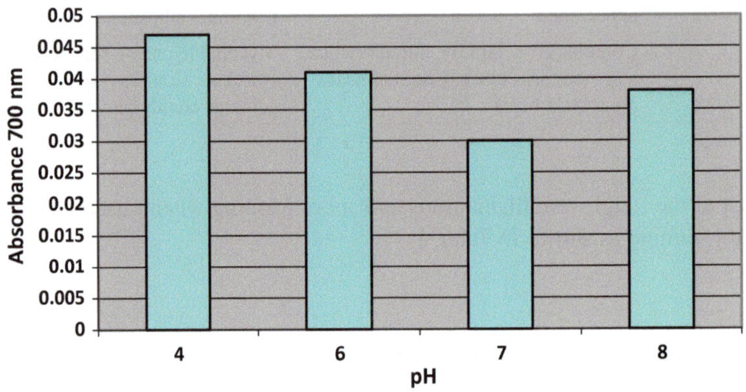

Fig. 3 The effect of pH on degradation activity of *Penicillium*

Figure 4, showed the effect of temperature on the ability of fungus to degrade the oil, it showed that extreme high temperature does not promote good degradation activity, the optimum Range was 30–55 °C and Maximum degradation activity was observed at 55 °C.

Figure 5 show that the Maximum degradation activity was seen at lowest concentration of oil 10%. High concentrations of oil may be exerting toxic effects on organisms.

As shown in Fig. 6; the highest degradation activity was seen at lowest concentration of NaCl, 0.5%.

Degradation of Pentadecane (C15) Nonadecane (C19) by Penicillium using thin layer chromatography were shown in Figs. 7 and 8.

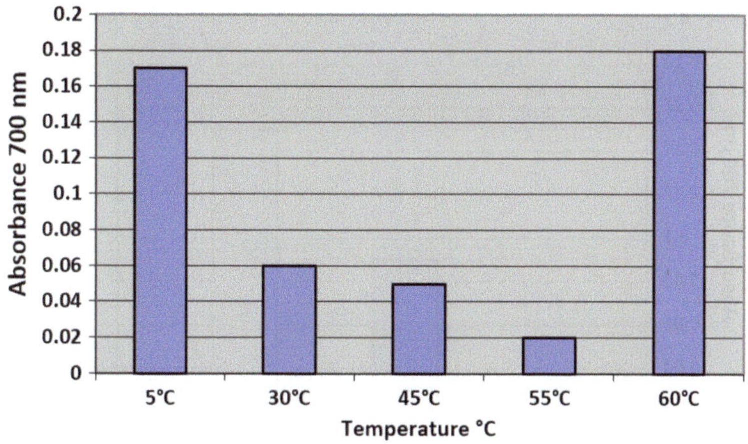

Fig. 4 The effect of temperature on degradation activity of *Penicillium*

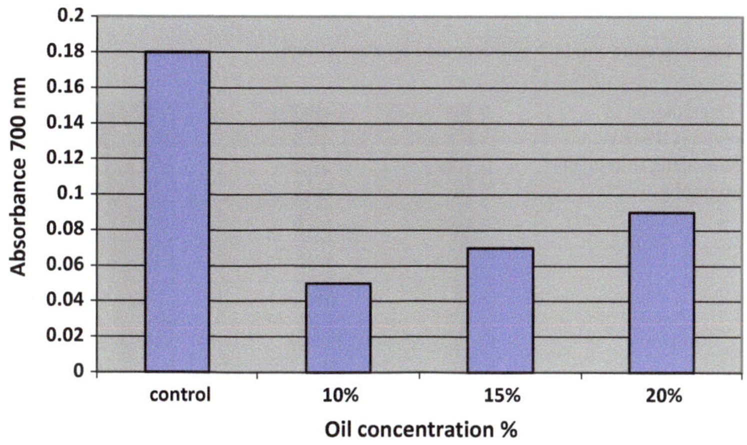

Fig. 5 The concentration of oil degraded by *Penicillium*

It can be seen that after optimizing various growth conditions, FG can efficiently degrade oil sample in 8 days in shaker incubator at 70 rpm when kept at 55 °C and pH 7 as shown in Fig. 9.

3.4 Degradation Studies

Checking degrading efficiency of different types of oil using Penicillium and consortium of Penicillium and Geobacillus (Absorbance at 700 nm) as shown in Table 2.

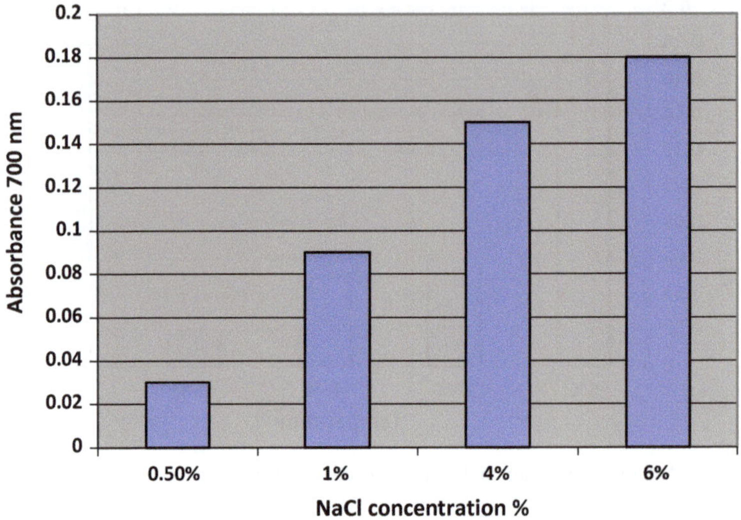

Fig. 6 The concentration of NaCl degraded by *Penicillium*

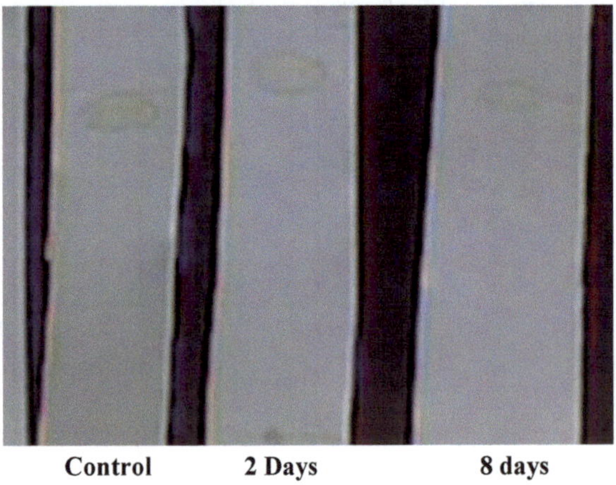

Fig. 7 Degradation of Pentadecane (**C15**) by *Penicillium* using thin layer chromatography

Table 2 Degrading efficiency of oils types

	Control	Fungus	Consortium
Paraffin	0.28	0.09	0.26
Nonadecane	0.40	0.05	0.35
Pentadecane	0.30	0.13	0.24
Sample oil	0.18	0.01	0.11

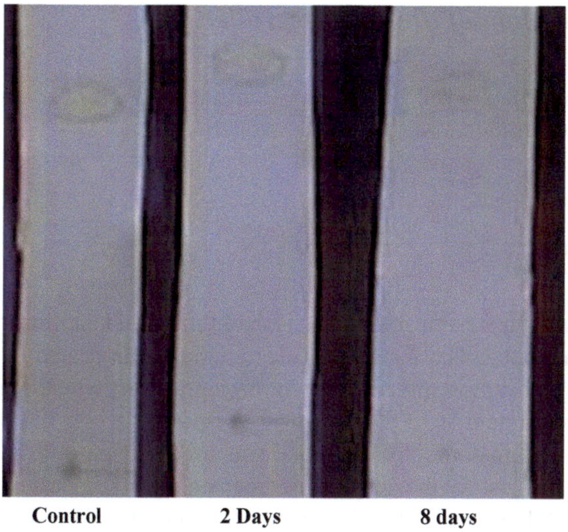

Fig. 8 Degradation of Nonadecane (**C19**) by *Penicillium* using thin layer chromatography

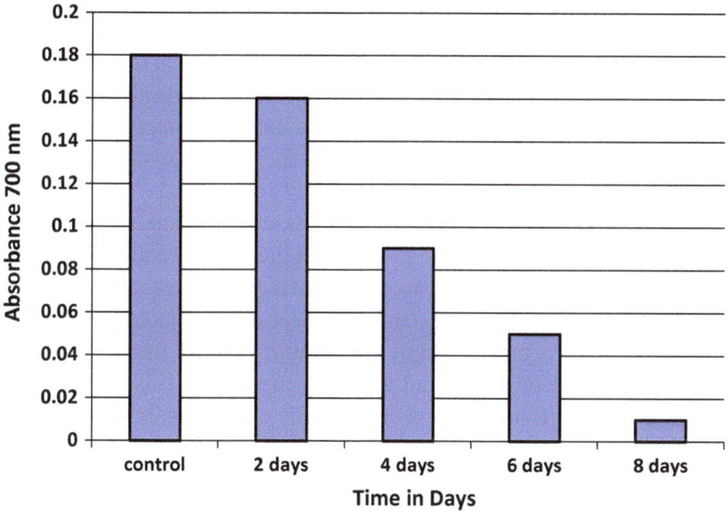

Fig. 9 Degradation pattern of the best degrader

3.5 *Identification of the Best Degrader*

Based on the morphological characterization, the best oil degrading fungus was identified as *Penicillium expansum* as shown in Fig. 9.

4 Discussion

Among our result the best degrader was isolated fungus Penicillium which showed the lowest absorbance. The measure of degradation is indicated by a lightness in color (as the hydrocarbon molecules were broken down) which is equivalent to a reduction in optical density [20].

Thin Layer Chromatography technique was used to measure the activity of the degradation on the basis of the decrease in the intensity of the crude oil. The reduced intensity on the 8th day indicated the degradation activity of Penicillium.

The effect of different parameter such as temperature show that at low temperatures, viscosity of the oil increases, and the volatilization of the toxic short chain alkanes is reduced and their water solubility is increased, delaying the onset of biodegradation [9]. There is a relation between the rates of hydrocarbon metabolism and temperature, the rate increased and may become maximum metabolism at higher temperature the range of 30–60 °C, above which the membrane toxicity of hydrocarbon increases [21]. Increase in temperature also influences increasing the bioavailability of the oil to the oil degrading organism and increases the oxidation rate by dissolving more oxygen into the medium [22].

Growth at 55 °C suggests that Penicillium, a mesophilic fungus could tolerate this temperature, the growth may be due to acquired thermotolerance. Fungi synthesize HSPs (heat shock proteins like protease) and acquire thermotolerance in the stressful conditions [23]. However, they have a transient heat shock response that is shortlived, even if they are able to grow at high temperatures. Most fungi favor a pH near neutrality, but fungi are more tolerant to acidic conditions. It is generally accepted that most of the fungi grow best in slightly acidic conditions in laboratory cultures [24, 25].

At the beginning of this study, the oil degrading bacteria and fungi were isolated when the pH of the soil sample was 6.8., i.e. almost near neutrality. This suggests that neutral conditions favor their degradation activity.

These results are consistent with one of the previous report, in which the effect of initial culture pH on phenanthrene degradation was examined on *Trametes versicolor*. The highest degradation occurred at pH 6. Neutral pH conditions were favorable for phenanthrene degradation by *Trametes versicolor* in this study [26].

Reduction of growth of fungus with the increasing concentration of oil may be because the high concentration of hydrocarbon can cause inhibition of biodegradation by nutrient or oxygen limitation or through toxic effects exerted by volatile hydrocarbons [27]. The removal percentage of phenanthrene by *T. versicolor* was higher at lower concentration of phenanthrene, and it is likely due to the toxicity of aromatic hydrocarbons. The removal percentage was highest at 10 mg/L of phenanthrene (76.7%), and lowest at 100 mg/L (59.1%). Han et al. [26], and in our research study the lowest concentration of oil (10%) is the maximum degrade activity. Presence of sodium chloride affects biosurfactant production which is very important in hydrocarbon degradation [28]. Penicillium showed degradation with in range of 0.5–4% NaCl. Optimal degradation was observed in 0.5% NaCl. So, NaCl concentration in medium affected cell growth and biosurfactant production. This indicates that surface tension reduced in 0–4% concentration of NaCl but lowest surface tension was at 0.5% concentration. This results are comparable with the results obtained by Lin and his colleagues in 1993 [29]. They showed that the addition of more than 4% NaCl in medium resulted in decrease in biosurfactant production. Degradation activity was observed in paraffin oil (C16–C22), which indicates that Penicillium is able to degrade paraffin wax. Most of the degradation was observed in test oil sample. It has been reported previously that *Penicillium funiculosum* could efficiently degrade the aliphatic carbons ranging from C12 to C30 [14].

5 Conclusion

The oil degrading fungi obtained by enrichment of the soil contaminated with oil in the mineral media was identified as *Penicillium expansum*. Thus, it can degrade oil with maximum efficiency at 55 °C, at around neutral pH, in presence of low concentrations of oil and NaCl, i.e. 10 and 0.5% respectively. Hence it can be effectively used to clear blockage problems in oil pipelines (in situ).

References

1. Bai, Y., Bai, Q.: Wax and asphaltenes. In: Subsea Pipelines and Risers, pp. 384. Elsevier (2005)
2. Leahy, J.G., Colwell, R.R.: Microbial degradation of hydrocarbons in the environment. Microbiol. Rev. **54**(3), 305–315 (1990)
3. Ezra, S., Feinstein, S., Pelly, I., Bauman, D., Miloslavsky, I.: Weathering of fuel oil spill on the east Mediterranean coast, Ashdod, Israel. Org. Geochem. **31**(12), 1733–1741 (2000)
4. Aitken, C.M., Jones, D.M., Larter, S.: Anaerobic hydrocarbon biodegradation in deep subsurface oil reservoirs. Nature **431**(7006), 291 (2004)
5. Bühler, M., Schindler, J.: Aliphatic hydrocarbons. In: Kieslich, K. (ed) Biotechnology, vol. 6, p. 56 (1984)
6. Spormann, A.M., Widdel, F.: Metabolism of alkylbenzenes, alkanes, and other hydrocarbons in anaerobic bacteria. Biodegradation **11**(2–3), 85–105 (2000)
7. Subramanian, D.T.: Microbial degradation of organic compounds. In: Gibson, D.T. (ed.) Microbiology Series, vol. 13. M. Dekker, New York (1984)
8. Thomas, K.: Heavy metals in urban fungi. Mycologist **6**(4), 195–196 (1992)
9. Atlas, R., Bartha, R.: Biodegradation of petroleum in seawater at low temperatures. Can. J. Microbiol. **18**(12), 1851–1855 (1972)

10. Margesin, R., Schinner, F.: Biodegradation and bioremediation of hydrocarbons in extreme environments. Appl. Microbiol. Biotechnol. **56**(5–6), 650–663 (2001)
11. Bonaventura, C., Johnson, F.M.: Healthy environments for healthy people: bioremediation today and tomorrow. Environ. Health Perspect. **105**(suppl 1), 5–20 (1997)
12. Kebria, D.Y., Khodadadi, A., Ganjidoust, H., Badkoubi, A., Amoozegar, M.: Isolation and characterization of a novel native *Bacillus* strain capable of degrading diesel fuel. Int. J. Environ. Sci. Technol. **6**(3), 435–442 (2009)
13. Liliana, J., Solórzano, L., Rizzo, C., Valeria, S., Soriano, A.U., Moura, M.I., Santos R.: Petroleum degradation by filamentous fungi. Paper presented at the a 9th international petroleum conference, EUA, Novo Mexico (2015)
14. Mancera-Lopez, M., Rodriguez-Casasola, M., Rios-Leal, E., Esparza-Garcia, F., Chavez-Gomez, B., Rodriguez-Vazquez, R., Barrera-Cortes, J.: Fungi and bacteria isolated from two highly polluted soils for hydrocarbon degradation. Acta Chim. Slov. **54**(1) (2007)
15. Oudot, J., Dupont, J., Haloui, S., Roquebert, M.: Biodegradation potential of hydrocarbon-assimilating tropical fungi. Soil Biol. Biochem. **25**(9), 1167–1173 (1993)
16. Sood, N., Lal, B.: Isolation of a novel yeast strain *Candida digboiensis* TERI ASN6 capable of degrading petroleum hydrocarbons in acidic conditions. J. Environ. Manage. **90**(5), 1728–1736 (2009)
17. Iheanacho, C.C., Okerentugba, P.O., Orji, F.A., Ataikiru, T.L.:Hydrocarbon degradation potentials of indigenous fungal isolates from a petroleum hydrocarbon contaminated soil in Sakpenwa community, Niger Delta. J. Phys. Nat. Sci. **3**(1), 6 (2014)
18. Haley, L., Trandel, J., Coyle, M.: Cumitech 11, practical methods for culture and identification of fungi in the clinical microbiology laboratory. In: Sherris, J.C. (Coordinating ed.). American Society for Microbiology, Washington, DC (1980)
19. Leck, A.: Preparation of lactophenol cotton blue slide mounts. Commun. Eye Health **12**(30), 24 (1999)
20. Adekunle, A., Adebambo, O.: Petroleum hydrocarbon utilization by fungi isolated from *Detarium senegalense* (J. F Gmelin) seeds. J. Am. Sci. **3**(1), 69–76 (2007)
21. Bossert, I., Bartha, R.: The fat of petroleum in soil ecosystems. In: Atlas, R.M. (ed.) Petroleum Microbiology, p.440–445. Macmillan, New York (1984)
22. Abdel-Fattah, Y.R., Hussein, H.M.: Numerical modelling of petroleum oil bioremediation by a local *Penicillium* isolate as affected with culture conditions: application of Plackett-Burman design. Arab. J. Biotechnol. **5**(2), 165–172 (2002)
23. Adams, P., Deploey, J.: Enzymes produced by thermophilic fungi. Mycologia **70**(4), 906–910 (1978)
24. Atlas, R.M.: Microbiology—Fundamentals and Applications. Macmillan Publishing Co., New York (1988)
25. Trent, J., Osipiuk, J., Pinkau, T.: Acquired thermotolerance and heat shock in the extremely thermophilic archaebacterium *Sulfolobus* sp. strain B12. J. Bacteriol. **172**(3), 1478–1484 (1990)
26. Han, M.-J., Choi, H.-T., Song, H.-G.: Degradation of phenanthrene by *Trametes versicolor* and its laccase. J. Microbiol. **42**(2), 94–98 (2004)
27. Leahy, J.G., Colwell, R.R.: Microbial degradation of hydrocarbons in environment. Am. Soc. Microbiol. Microbiol. Rev. **54**(3), 16 (1990)
28. Rismani, E., Fooladi, J., Ebrahimi Por, G.H.: Biosurfactant production in batch culture by *Bacillus licheniformis* isolated from the Persian Gulf. Pak. J. Biol. Sci. **9**(13), 5 (2006)
29. Lin, S.C., Sharma, M.M., Georgiou, G.: Production and deactivation of biosurfactant by *Bacillus licheniformis* JF-2. Biotechnol. Prog. **9**(2), 138–145 (1993)

Gastrointestinal Larval Nematodes on Pastures Grazed by Small Ruminants of Duhok Area

Adnan M. Abdullah Al-Rekani, Ronak Abdulaziz Meshahbaz, Abdulrehman Abdulhamid Yousif and Fatah Majeed Khalaf

Abstract The aim of the present study was determining the contamination and prevalence of gastrointestinal (GI) nematodes larvae on pastures permanently grazed by sheep and goats in different areas of Duhok province under the climate of the spring season. Samples of herbage were collected from different pastures and the larvae were identified and then counted. Out of 144 herbage samples, 89 (61.81%) were found positive in all selected areas during the three months (March, April and May). Significant differences ($p < 0.05$) in contamination rates and pasture larvae count were observed among the months of study whereas they did not significant among the study areas. The most favorable environmental conditions for survival and development of nematodes larvae on pasture were found in April and followed by March. Also, a significant correlation was observed between environmental variables and pasture larvae three-stage count (PL_3C). It can be concluded that changes in environmental conditions among the months of spring season have a significant effect on the contamination of GI nematodes larvae in different grazing pastures of Duhok province. These results could be beneficial in planning control program of nematodes parasites.

Keywords Contamination · Nematode · Larvae · Pasture · Sheep and goats

---

A. M. Abdullah Al-Rekani (✉) · A. Abdulhamid Yousif · F. Majeed Khalaf
Department of Animal Production, College of Agriculture, Duhok University, Duhok, Iraq
e-mail: adnanrekani@uod.ac

A. Abdulhamid Yousif
e-mail: abdulrahmanayousif@gmail.com

F. Majeed Khalaf
e-mail: fatah.khalaf2112@gmail.com

R. Abdulaziz Meshahbaz
Department of Environmental Science, College of Science, Zakho University, Zakho, Iraq
e-mail: ronak.meshabaz@uoz.edu.krd

Y. T. Mustafa et al. (eds.), *Recent Researches in Earth and Environmental Sciences*,
Springer Proceedings in Earth and Environmental Sciences,
https://doi.org/10.1007/978-3-030-18641-8_12

1 Introduction

Parasitic nematodes infection act as mainly a cause of losses of economic in the production of small ruminant in the world [1]. In developing countries, gastrointestinal (GI) nematodes are assessed as a main constraint of the ruminants production [2]. As a result of these nematodes severe pathological impacts, they hold high importance [3]. Infection with GI nematodes is considered to a great extent to be extremely important disease of grazing goats and sheep throughout the world, leading to reduction of productive performance, parasitic gastroenteritis and even mortal [4, 5].

The outbreak of gastrointestinal nematodes infections that is associated with the agro-climatic conditions such as pasture's quantity and quality, ambient temperature, relative humidity, rainfall, stocking rate and animal's grazing behavior [6]. Moreover, the major source for nematode infection remains the pastures contaminated with nematodes infective larvae, and this contamination occurs from infected animals coming for grazing. A susceptible animal becomes infected by ingestion of GI nematode infective larvae along with contaminated pastures [7].

The survival and development of free living stages of GI nematodes of sheep and goats are affected generally by humidity and ambient temperature with the influences of pasture conditions acting a major role [8]. The favorite development condition from egg to the L3 is 18–26 °C with relative humidity more than 70% [9, 10]. In addition; seasonal changes in the environment causes a transmission rates of GI nematodes larvae by affecting the survival and the development of larvae in the environment and host contact with infectious free living stages [9]. A study of Al-Dabagh [11] also noticed that the suitable seasons of eggs' hatching and larval stages' development were in spring and autumn under Iraq's climate condition. The geo-climate condition of Iraq during the spring season is crucial for the optimal growth and proliferation of GI nematode's larvae.

Estimation of infective nematodes larvae in the pasture indicates an infection in animals grazed in particular pastures and it provides important information and insight in the population dynamics of several GI nematodes species with free-living stages [12]. Information related to the pasture contamination with nematode infective larvae in Duhok province is limited, therefore the aim of the present study was determining the contamination and prevalence of infective larvae of GI nematode on sheep and goat grazing pasture in different areas of Duhok province under the climate of the spring season. This research is considered the first study of pasture contamination with GI nematode larvae over Duhok province, which is located in Iraqi Kurdistan region and lies between longitude (42°–45° E) and latitude (36°–38° N).

2 Materials and Methods

2.1 Study Area and Collection of Herbage Samples

Total of 144 herbage samples was collected from four areas in Duhok Province (surrounding areas of Duhok, Sumel, Bagera, and Zakho city) during three months of the spring season (March, April, and May of 2015). From each area, three grazing pastures of sheep and goats were randomly selected as sampling areas, and from each pasture, four replicate herbage samples approximately about 250–300 g of weight were collected at monthly interval. Each herbage sample was collected while walking across the pasture in a W-shape pattern and cut using scissors at the plant's base, without including soil or roots as describe by Stadalienė et al. [13]. Each sample placed in clean plastic cistern and transferred into the laboratory of Animal Production Department at the College of Agriculture/Duhok University for separation and examination of nematodes larvae.

The number of herbage samples per month = 4 replicates × 3 pastures × 4 areas = 48 samples.

2.2 Separation of Larvae from Herbage

Larvae of GI nematodes were recovered and isolated from the samples' of herbage based on the technique of Fernández et al. [14] and Hansen and Perry [15]. Samples were put into a container filled with water thereafter a small amount of detergent (liquid soap) was added to water and the mixture was agitated by hand for facilitating separation of larvae from the herbage and left during one day. Afterwards the herbage was removed from the containers and the water was poured onto the strainer, wherever infective larvae were retained. The water with larvae then transferred into a Baermann's funnel at room temperature, and the number of larvae were recovered on the next time and then counted.

2.3 Identification of Larvae

The larvae of nematode were identified under a light and digital microscopic (Dinoeye) based on the morphological characteristics like length of larvae, Length of sheath tail extension, number and shape of intestinal cells as show in Appendices (1, 2) and (Figs. 2 and 3). The measurements of these morphological characteristics were estimated and identified according to descriptions and techniques provided by Van Wyk and Mayhew [16], Großbritannien [17] and Urquhart et al. [18]. Herbage larvae count were calculated as the number of larvae per kilogram of dry herbage (PL_3C/kg DM). PL_3C/kg DM = Count × 1000/Herbage weight.

2.4 *Climatic Data*

Climatic data such as temperature, relative humidity and rainfall during the study period were gained at meteorological stations located in the college of agriculture in Sumel area, Duhok and Zakho city.

2.5 *Statistical Analysis*

Collected data were statistically tabulated by descriptive statistics and analyzed by ANOVA: two-factor with replication analysis of variance (two-way ANOVA) and correlation coefficient using Microsoft Excel 2007 and SPSS 18.0 software and differences were considered significant at $P < 0.05$.

3 Result

During three months of the spring season, 144 herbage samples from selected grazing pastures in four areas of Duhok province were collected and examined to detect the prevalence of GI nematodes larvae. The result of this study revealed that 89 (61.81%) out of 144 herbage samples were found positive for the nematode larvae as indicated in Table 1. The differences in contamination rates and their percentages among the months were statistically significant ($p < 0.05$) while among the selected study areas were not. The highest percentage of larval contamination in all study areas was recorded in April (83.33%) and followed by March (68.75%) while in May was declined to the lowest level (33.33%).

Figure 1 shows monthly pasture larvae count in four study areas of Duhok province. Statistically, there were significant differences ($p < 0.05$) among months while non-significant differences among study areas. The high mean of pasture lar-

Table 1 Areas and months-wise GI nematodes larvae contamination of Duhok province

Positive samples/examined samples*				
Areas	March (%)	April (%)	May (%)	Total (%)
Duhok	7/12* (58.33)	10/12 (83.33)	4/12 (33.33)	21/36 (58.33)
Bagera	9/12 (75)	11/12 (91.67)	1/12 (8.33)	21/36 (58.33)
Sumel	9/12 (75)	11/12 (91.67)	8/12 (66.67)	28/36 (77.78)
Zakho	8/12 (66.67)	8/12 (66.67)	3/12 (25)	19/36 (52.78)
Total mean	33/48 (68.75)[a]	40/48 (83.33)[b]	16/48 (33.33)[c]	89/144 (61.81)

[a,b,c]Means with different small letters in the same row differ significantly
*$p < 0.05$

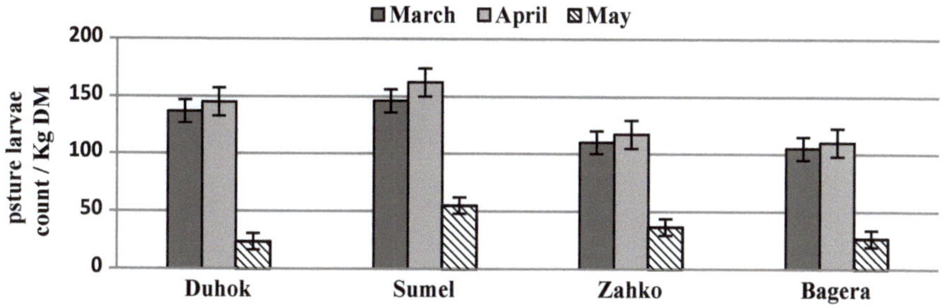

Fig. 1 Monthly mean of pasture larval counts in different districts of Duhok province PL_3C/Kg DM (kg DM = kilogram of dry matter [herbage])

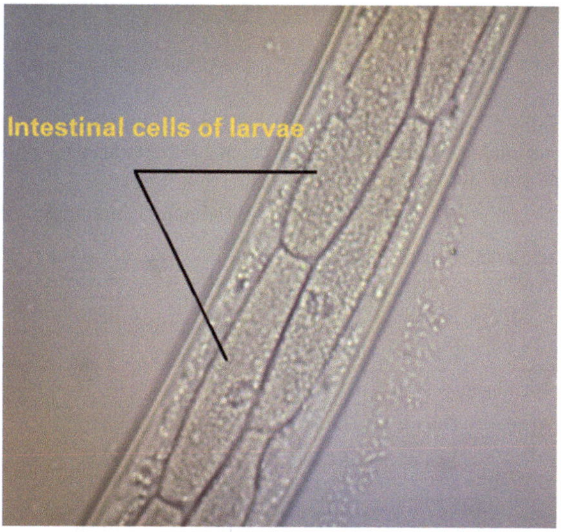

Fig. 2 Intestinal cells of *Chabertia ovina* infective larva (L_3)

vae count was recorded in April following by March, while it was declined to the lowest level in May.

Larvae of *Ostertagia* spp., *Trichostrongylus* spp., *Haemonchus contortus*, *Nematodirus* spp. and *Chabertia ovina*, in varying numbers, were recovered from herbage samples. Figures 2 and 3 illustrate the intestinal cells of infective larva of *Chabertia ovina* and *Trichostrongylus* spp. respectively. There is a great difference in the shape of the intestinal cells between these species of nematodes larvae. The intestinal cells of *Chabertia ovina* larva appear as shape of regular rectangles while in *Trichostrongylus* spp. larva they appear as shape of triangles.

Table 2 shows the climatic conditions as (maximum temperature, minimum temperature, humidity, and rainfall) during the months of spring (March, April and May) with contamination percentage and count of pasture larvae in all selected areas. The highest percentage of positive samples of contamination larvae and mean pasture

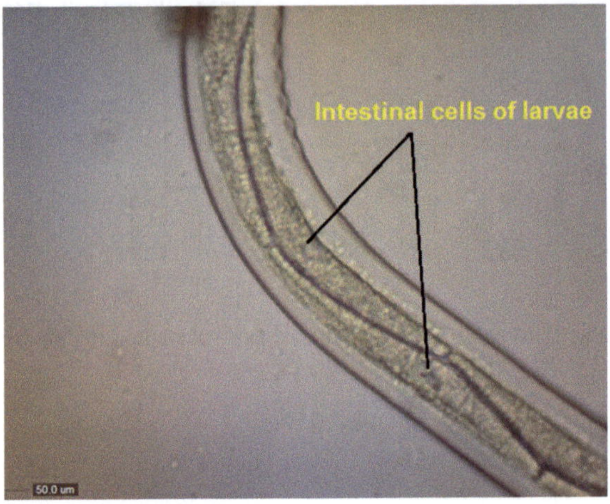

Fig. 3 Intestinal cells of infective larva (L_3) *Trichostrongylus* spp.

Table 2 Mean of environmental elements (maximum and minimum temperature, relative humidity and rainfall) with pasture larvae count and contamination percentage of different areas of Duhok province during spring season

Month	Districts	Maximum tem. (°C)	Minimum tem. (°C)	Humidity (%)	Rainfall (mm)	PL_3C/kg DM	CR (%)
March	Dohuk	17.55	8.15	69.48	71.40	136.67	58.33
	Sumel	18.21	6.15	70.20	47.80	146.33	75.00
	Zakho	18.56	8.01	62.93	67.20	110	66.67
Total mean		18.11	7.44	67.54	62.13	131	66.67
April	Dohuk	22.73	11.61	66.87	40.20	145	83.33
	Sumel	24.03	8.49	59.00	22.20	162.66	91.67
	Zakho	23.00	11.90	53.70	32.50	117	66.67
Total mean		23.25	10.67	59.86	31.63	141.22	80.56
May	Dohuk	30.81	18.18	46.68	9.60	23.33	33.33
	Sumel	33.07	14.90	37.30	11.40	55	66.67
	Zakho	31.99	17.54	37.83	3.30	37	25.00
Total mean		30.81	16.87	40.60	8.10	38.33	41.67

PL_3C/kg DM = Pasture infective larvae count per kg dry matter
Tem. = Temperature
CR % = Percentage of the contamination rate

larval count are present under the climate of April and followed by March in all study areas, while the lowest levels are shown under the climate of May.

Correlations of environmental condition (temperature, relative humidity, and rainfall) with pasture larval count (PL_3C) are shown in the Table 3. Statistically, there was a significant negative correlation between the pasture larval count and each of

Table 3 Correlation coefficient between the environmental variables with pasture infective larvae count

Environmental variables	PL_3C
Maximum tem. (°C)	−0.82*
Minimum tem. (°C)	−0.90*
Relative humidity (%)	0.85*
Rainfall (mm)	0.65*

*Significant correlation at (0.05) level
PL_3C = Pasture infective larvae count

the maximum and minimum temperature while the correlation between the pasture larval count and each of the relative humidity and rainfall was significantly positive.

4 Discussion

The present study was conducted to investigate the contamination of pasture with GI nematode larvae in Duhok province during spring season (March, April, and May). The larvae of several species of GI nematodes were observed in this study and these were *Ostertagia* spp., *Trichostrongylus* spp., *Haemonchus contortus* and *Nematodirus* spp. and *Chabertia ovina*. All these species except *Chabertia ovina* were recorded in previous epidemiological studies obtained by several researchers in other regions of Iraq [19–22] in Duhok governorate. While the species of *Chabertia ovina* detected in some pastures of Duhok province had not been observed by previous studies in Duhok area. The shape of intestinal cells of *Chabertia ovina* larvae greatly differs with *Trichostrongylus* spp. larvae as indicate in (Figs. 2 and 3). The shape of these intestinal cells could be useful for diagnosis and identification of infective larvae of some nematodes species.

The results indicated that there was a marked variation in contamination rate and pasture larvae count among months over the study period. The highest percentage of availability of infective nematodes larvae and mean of pasture larvae count in different study areas were recorded in April and followed by March. This high prevalence may be due to the environmental conditions of April and May that are favorable for development and survival of nematodes larvae on pasture and the poor control measures. The contamination rate and count of nematodes larvae were significantly reduced in May could be attributed to the beginning of dry conditions and the end of green pasture period. These variations in pasture larvae count and contamination rate among months are in agreement with findings of Al-Shaibani et al. [23], Gadahi et al. [24] and Das et al. [25]. O'Connor et al. [8] viewed that the highest level of pasture contamination with free-living infective larvae (L3) during rainy season, while lower contamination was found during dry season.

The changes of maximum and minimum ambient temperature were demonstrated the major factors governing the variation of pasture larvae count and contamination rate among the months. And the optimum climatic conditions for development and survival of GI nematodes larvae during the period of study were observed in the month of April, where the maximum and minimum temperature were 23.2 and 10.6 °C respectively, with relative humidity 59.8% as total means of all study areas. These results coincide with those found by Rossanigo and Gruner [9] and Nahar et al. [10].

Statistical analysis showed a significant negative correlation between pasture larvae three-stage count (PL_3C) and each of maximum and minimum temperature, and a significant positive correlation between pasture larvae count and each of relative humidity and rainfall (Table 3) that is in agreement with results of Gadahi et al. [24] and Agyei et al. [26]. The negative correlation between ambient temperature and PL_3C may be caused by an increase in temperature that was observed in the month of May. The growth of the nematode eggs to infective larvae and the subsistence of these larvae on pasture are influenced by many factors and the most important are environmental conditions that include ambient temperature, relative humidity and rainfall [7]. A study of O'Connor et al. [8] suggested that eggs deposited in the soil to reach the infective larvae (L3) are mainly effected by temperature and humidity. Environmental elements are important not only for the development and survival of larvae stage but also for the movement of the infective larvae to the herbage. Khadijah et al. [27] shows that the development of *Haemonchus contortus* and *Trichostrongylus colubriformis* free living stage to the third stage larvae (L3) highly influenced by soil moisture and its interaction with rainfall. Also, Agyei and Amponsah [28] reported that significant positive relationship exists between the level of infective larvae on herbage and rainfall. Studies carried out on nematode larvae ecology in some parts of sub-Saharan Africa [29], Europe, Australia and the Pacific islands [30] have observed that the proportion of development and the permanency of eggs vary at different temperatures and humidity and in different geo-ecological areas. Also, free water was necessary for larvae to migrate from the dung, but not for vertical movement onto plant [31].

Other factors related to prevalence and contamination of nematodes larvae are ecological zone, husbandry practices, management systems, types of grass, species of helminths, stoking rate of animals and physiological status of the animal [32–34]. Worku, and G. Dinede, indicated that species of nematodes are very fertile in which a single female nematode can produce many thousands of eggs (or larvae) each day. The epidemiological significance of contaminated pastures further raises under traditional small ruminant husbandry system, communal grazing and in the absence of strategic deworming programs in developing countries [7].

5 Conclusion

According to results of this study, the changes of environmental factors (temperature, humidity and rainfall) among the months of spring season have a significant effect on the development and survival of free-living stages of GI nematodes in the selected pasture. The most favorable conditions related to infective larvae development were in the month of April, followed by March. The grazing pastures of sheep and goats are remained contaminated with the GI nematodes larvae during these favorable months. Therefore, planning of pasture management strategies is important to assist in the control of gastrointestinal nematodes through reducing of contamination and ingestion of infective larvae by animals. Also deworming of animals before they are released for grazing, particularly at the start of spring is suggested in order to keep the pasture free from nematodes larvae contamination.

Appendices

Appendix 1

See Fig. 4.

Appendix 2

See Fig. 5.

Fig. 4 Picture of infective larva (L_3) of *Trichostrongylus* spp.

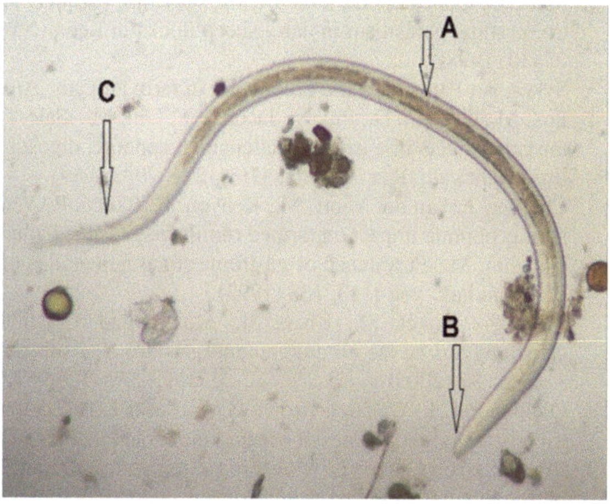

A= intestinal cells, B=head of larva, C= tail of larva

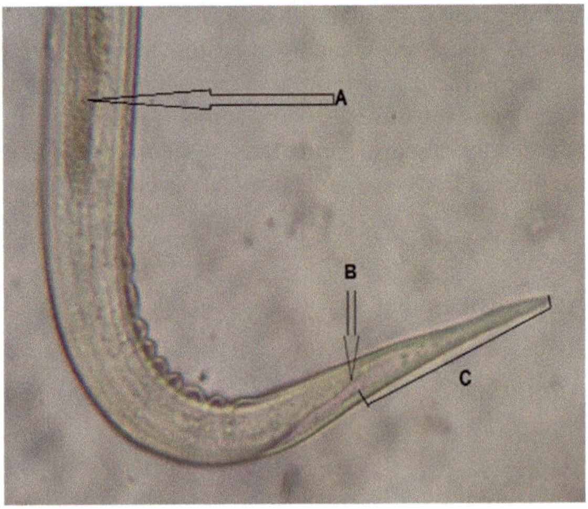

A= intestinal cell, B= tip of tail, C= sheath tail extension

Fig. 5 Some morphological characteristics of infective larva (L_3) tail of *Trichostrongylus* spp.

References

1. Suarez, V., Busetti, M.: The epidemiology of helminth infections of growing sheep in Argentina's western pampas. Int. J. Parasitol. **25**(4), 489–494 (1995)
2. Martínez-González, B., Díez-Baños, N., Rojo-Vázquez, F.: An epidemiological study of gastrointestinal parasitism in dairy sheep flocks in León (NW Spain). Small Rumin. Res. **27**(1), 25–30 (1998)
3. Sykes, A.: Parasitism and production in farm animals. Anim. Sci. **59**(2), 155–172 (1994)
4. Fox, M., Gerrelli, D., Pitt, S., Jacobs, D., Gill, M., Gale, D.: *Ostertagia ostertagi* infection in the calf: effects of a trickle challenge on appetite, digestibility, rate of passage of digesta and liveweight gain. Res. Vet. Sci. **47**(3), 294–298 (1989)
5. Charlier, J., van der Voort, M., Kenyon, F., Skuce, P., Vercruysse, J.: Chasing helminths and their economic impact on farmed ruminants. Trends Parasitol. **30**(7), 361–367 (2014)
6. Qayyum, M.: Prevalence of gastrointestinal nematodes of sheep and goats in upper Punjab, Pakistan. Pak. Vet. J. **13**, 138 (1993)
7. Iqbal, Z., Lateef, M., Khan, M., Muhammad, G., Jabbar, A.: Temporal density of trichostrongylid larvae on a communal pasture in a sub-tropical region of Pakistan. Pak. Vet. J. **25**(2), 87 (2005)
8. O'Connor, L.J., Walkden-Brown, S.W., Kahn, L.P.: Ecology of the free-living stages of major trichostrongylid parasites of sheep. Vet. Parasitol. **142**(1–2), 1–15 (2006)
9. Rossanigo, C., Gruner, L.: Moisture and temperature requirements in faeces for the development of free-living stages of gastrointestinal nematodes of sheep, cattle and deer. J. Helminthol. **69**(4), 357–362 (1995)
10. Nahar, L., Sarder, M., Mondal, M., Faruque, M., Rahman, M.: Prevalence of haemonchosis of goats at Rajshahi district in Bangladesh. Bangladesh J. Vet. Med. **13**(1), 29–36 (2015)
11. Al-Dabagh, A.T.M.: Field study on the bionomic of ovine Abomasum nematode *Haemonchus contortus* (in Iraq) (1984)
12. Grenfell, B., Smith, G., Anderson, R.: A mathematical model of the population biology of *Ostertagia ostertagi* in calves and yearlings. Parasitology **95**(2), 389–406 (1987)
13. Stadalienė, I., Höglund, J., Petkevičius, S.: Seasonal patterns of gastrointestinal nematode infection in goats on two Lithuanian farms. Acta Vet. Scand. **57**(1), 16 (2015)

14. Fernández, S., Šarkunas, M., Roepstorff, A.: Survival of infective *Ostertagia ostertagi* larvae on pasture plots under different simulated grazing conditions. Vet. Parasitol. **96**(4), 291–299 (2001)
15. Hansen, J., Perry, B.: The Epidemiology, Diagnosis and Control of Helminth Parasites of Ruminants, 2 (1994)
16. Van Wyk, J.A., Mayhew, E.: Morphological identification of parasitic nematode infective larvae of small ruminants and cattle: a practical lab guide. Onderstepoort J. Vet. Res. **80**(1) (2013)
17. Großbritannien, M.o.A.: Manual of Veterinary Parasitological Laboratory Techniques, 160 p., Ill. HM Stationery Office (1986)
18. Urquhart, G.M., J.A., Duncan, J.L., Dunn, A.M., Jennings, F.W.: Veterinary parasitology. In: English Language Book Society (2000)
19. Al-Barwary, L.T.: A Comparative Study of Some Epidemiological Aspect of the Abomasal Nematodes of Sheep and Goats in Duhok Governorate-Iraq. Coll. of Vet. Med., University of Duhok
20. Al-Rekani, A.: Effect of natural infection with gastrointestinal nematode on milk composition and blood parameters of lactating native goats. Int. J. Sci. Res. **1**(2), 14–17 (2012)
21. Ali, W.R.: Gastrointestinal Nematodes in Iraqi Goats and Their Pathogenic Effect in Experimental Infected Animals. Coll. of Sci., Baghdad University (1985)
22. Al-Saeed, A., Al-Khalidi, N.: Epidemiological studies on abomasal nematodes of sheep in Mosul (Iraq). J. Vet. Parasitol. **4**(1), 17–20 (1990)
23. Al-Shaibani, I., Phulan, M., Arijo, A., Qureshi, T.: Contamination of infective larvae of gastrointestinal nematodes of sheep on communal pasture. Int. J. Agric. Biol. **10**(6), 653–657 (2008)
24. Gadahi, J.A., Bhutto, B., Gilgiti, N.A., Akhter, N., Arijo, A.G.: Contamination of larvae of gastrointestinal nematodes on animal grazing pastures in and around Tandojam, Pakistan. Int. J. Agro Vet. Med. Sci. **5**, 389–394 (2011)
25. Das, M., Deka, D., Islam, S., Sarmah, P., Bhattacharjee, K.: Gastrointestinal nematode larvae in the grazing land of cattle in Guwahati, Assam. Vet. World **9**(12), 1343 (2016)
26. Agyei, A., Aklaku, I., Djang-Forjour, K., Dodoo, R., Fynn, K.: Epidemiological studies on the gastrointestinal parasitic infection of lambs in the Guinea and transitional savanna regions of Ghana. Ghana J. Agric. Sci. (2005)
27. Khadijah, S., Kahn, L., Walkden-Brown, S., Bailey, J., Bowers, S.: Soil moisture influences the development of *Haemonchus contortus* and *Trichostrongylus colubriformis* to third stage larvae. Vet. Parasitol. **196**(1–2), 161–171 (2013)
28. Agyei, A., Amponsah, A.: Seasonal availability of infective strongylate nematode larvae on pasture in the forest zone of Ghana. Bull. Anim. Health Prod. Afr. **49**, 68–72 (2001)
29. Onyali, I., Onwuliri, C., Ajayi, J.: Development and survival of *Haemonchus contortus* larvae on pasture at Vom, Plateau State, Nigeria. Vet. Res. Commun. **14**(3), 211–216 (1990)
30. Besier, R., Dunsmore, J.: The ecology of *Haemonchus contortus* in a winter rainfall region in Australia: the development of eggs to infective larvae. Vet. Parasitol. **45**(3–4), 275–292 (1993)
31. Morgan, E.: The influence of water on the migration of infective trichostrongyloid larvae onto grass. Parasitology **138**(6), 780–788 (2011)
32. Assoku, R.: Studies of parasitic helminths of sheep and goats in Ghana. Bull. Anim. Health Prod. Afr. **29**(1), 1–10 (1981)
33. Mushonga, B., Habumugisha, D., Kandiwa, E., Madzingira, O., Samkange, A., Segwagwe, B.E., Jaja, I.F.: Prevalence of *Haemonchus contortus* infections in sheep and goats in Nyagatare District, Rwanda. J. Vet. Med. **2018** (2018)
34. Worku, H., Dinede, G.: Epidemiological study of nematodes in Adigudom and Mereme'eti grazing areas, southern Tigray, northern Ethiopia: larval nematodes cattle pasture contamination. Sci. Publ. J. **8**(9), 6 (2016)

[illegible]

MIX
Papier aus verantwortungsvollen Quellen
Paper from responsible sources
FSC® C105338

If you have any concerns about our products,
you can contact us on
ProductSafety@springernature.com

In case Publisher is established outside the EU,
the EU authorized representative is:
Springer Nature Customer Service Center GmbH
Europaplatz 3, 69115 Heidelberg, Germany

Printed by Libri Plureos GmbH
in Hamburg, Germany